IS SCIENCE MULTICULTURAL?

RACE, GENDER, AND SCIENCE

Anne Fausto-Sterling, *General Editor*

Is Science Multicultural?

POSTCOLONIALISMS, FEMINISMS, AND EPISTEMOLOGIES

Sandra Harding

INDIANA UNIVERSITY PRESS

Bloomington and Indianapolis

The paper used in this publication meets the minimum
requirements of American National Standard for Information
Sciences—Permanence of Paper for Printed Library
Materials, ANSI Z39.48-1984.

Manufactured in the United States of America

Library of Congress Cataloging-in-Publication Data

Harding, Sandra G.
    Is science multicultural? : postcolonialisms, feminisms,
and epistemologies / Sandra Harding.
       p.   cm. — (Race, gender, and science)
    Includes bibliographical references and index.
    ISBN 0-253-33365-2 (cloth : alk. paper). —
    ISBN 0-253-21156-5 (pbk. : alk. paper)
    1. Science—Social aspects. 2. Science—History.
3. Technology—History. 4. Feminism. 5. Decolonization.
I. Title. II. Series.
Q175.5.H39   1998
501—dc21   97-28249

1 2 3 4 5 03 02 01 00 99 98

CONTENTS

PREFACE

Is it true that we write for ourselves as much as we do for our imagined readers? The project of this study has been developing for more than fifteen years as I have tried to figure out how to map for myself and my students relationships between literatures and arguments that it has seemed to me would benefit from greater dialogue with each other.

This project began by reflecting on inadequacies in the conceptual frameworks of contemporary epistemology and philosophy of science from the perspective, on the one hand, of issues raised in the women's movement and feminist theory of the 1970s and early 1980s and, on the other hand, in the emerging African and African American philosophy of the period. In both cases, the purported universal validity and desirability of the prevailing standards of objectivity, rationality, and good research methods were being challenged by the new social movements and critiques in the disciplines.[1] Feminist epistemology and science studies began their immensely fruitful projects during this period, and my own thinking was greatly stretched by encounters and exchanges with scholars from other disciplines.[2] In my mind, somehow linking the feminist and beginnings of postcolonial work, though only in occasional and fragile interaction with them in fact, were the powerful political and literary writings of U.S. women of color. Clearly women were not a homogenous group, women's thinking and that of other cultures were not any more irrational or unobjective than was modern European/North American culture, and there was something deeply wrong with the disciplinary standards of good research in history, the natural and social sciences—and the philosophical assumptions that made them appear plausible to so many well-intentioned people.

During the same period, new approaches to science and technology studies began to flourish in Europe and North America, pursuing the call of the social historians to show the integrity of events and processes in the history of modern science with their historical eras, as Thomas Kuhn put the point in his influential book, *The Structure of Scientific Revolutions*.[3] The sociologists and ethnographers soon joined the historians in using the same analytic tools that they used to account for error and falsity to examine the processes through which information in the natural sciences became legitimated as knowledge and truth.[4] Scientific processes were and are social processes, of course, that both enhance and limit the role that nature can play in legitimating information as knowledge and truth. These historians, sociologists, and ethnographers were explicitly raising philosophic questions. Some thought epistemology and the philosophy of science inherently so arrogant and out of touch with the practices of sciences as to be useless at best, and others called for epistemologies capable of accounting for the new kinds of histories and sociologies of what all still agreed was the worthy ideal of human knowledge production—modern science. Philosophers joined them in these skeptical reflections on the status philosophy had claimed for itself and in calls for new philosophies.[5] Feminists and, as I would soon discover, postcolonial writers were thinking along similar lines and already beginning to produce epistemologies and philosophies of science the themes of which partly overlapped with those of the post-Kuhnian Euro-American science and technology studies.

There was little awareness in the post-Kuhnian literature, however, that the "historical eras" with which sciences have been integrated might have gender, race, and global imperial dimensions as well as the other cultural ones these authors had so brilliantly revealed. And the northern feminist epistemologies and science and technology studies, though they often mentioned ". . . and race," with notable exceptions did not seem to know how to think feminist epistemology, science, and technology issues onto the map that the new multicultural and global feminist writings were already producing.[6]

In fact, clues to such postcolonial science and technology maps were already emerging, but in institutional locations hardly visible to the northern post-Kuhnian and feminist thinkers. Various fortuitous events led me to them. In 1990 I lectured on feminist science and epistemology issues for a month in Costa Rica, Nicaragua, Honduras, and Guatemala sponsored by the United Nations' Pan American Health Organization's "Women, Health and Development" project and the University of Costa Rica. I had to ask myself why the science and technology questions centered in the northern feminist discourses were off on the side of the Central American concerns. The northern feminist issues were important to the PAHO organizers and to the national Ministry of Health directors with whom I met in these countries; but these scholars and policy analysts kept trying to work out their relevance for conditions for which the northern frameworks had not been designed. I then turned to development studies for clues to how to bridge such gaps. Though I had glanced through the "women in development" writings at various earlier times, these had seemed to me too mired in precisely the problematic older understandings of modern science and technology to enable one to envision more appropriate epistemologies and philosophies of science and technology. They had been calling for adding women to processes of development that themselves assumed the pre-Kuhnian epistemologies and philosophies of science and technology that, it seemed to me, fit all too well with distinctly bourgeois and imperial politics and economics—though I couldn't have spelled out how in much detail at that time. Philosophical "noblesse oblige," however, did not seem to me to be an appropriate or useful approach to science, technology, and knowledge issues in the so-called underdeveloped societies.

Now I took another look at the successor to this earlier literature and discovered that feminist critiques of economistic notions of development were emerging, with science and technology themes central to them. The field was in the process of moving from a focus on "women in development" to its current concerns with women, the environment, and alternatives to development. (Now referred to as WEAD.) At this time I also began seeking materials to add to the essays I had already been teaching in order to edit an anthology, The "Racial" Economy of Science: Toward a Democratic Future.[7] Through fortuitous encounters in development journals and bookstores and tips from colleagues, I began to enlarge the small collection of books and essays I had been teaching and thinking about that could be called a postcolonial science and technology literature.

In 1992, two events stimulated new studies and the wider dissemination of older ones that pursued these postcolonial science and technology themes. One was the United Nations' Rio Earth Summit, preceded by a feminist environmental conference in Miami. The other was the commemoration of the Columbian Quincentennial. A number of books, essays, and collections on the themes of these two events began to show the causal links between the emergence and flourishing of modern sciences in Europe, on the one hand, and European expansion, on the other hand—originally through the Voyages of Discovery and in recent decades through international development policies. And feminist themes found a place in at least some parts of both literatures. I began attending national and international conferences focussed on comparative science and technology traditions and on rethinking science and technology philosophy and policy for the future in multicultural and global contexts. Then the next year many science- and technology-focussed United Nations organizations began to prepare for the United Nations Fourth International Conference on Women, to meet in Beijing in 1995, and I had the opportunity to work with the Gender Working Group of the U.N. Committee on Science and Technology for Development and with the United Nations Educational, Scientific, and Cultural Organization's (UNESCO) World Science Report 1996. These gave me the opportunity to work with science, technology, and feminist issues in the most expansive global frameworks and to try to grasp more clearly relations between philosophies and such arenas of science and technology policies.[8]

And so the mappings that this book pursues took shape in response to these interests and fortuitous encounters during the historical changes of the last decade and a half. They reflect the idiosyncrasies of my particular history of attempts to think more accurately, comprehensively, and usefully about philosophic issues that have interested me since graduate school, and that have turned out to be of concern to a lot of other people thinking about sciences, technologies, knowledge, and power in the multicultural and postcolonial world in which we all find ourselves. Others do and would think their way through these issues differently and center other issues of greater concern to them. This intends only to be *a* mapping of a terrain that has interested me and some others—not *the* mapping of it. Indeed, the necessity of developing philosophies that claim both far less and far more than only universal validity is one of the themes to be pursued here. This map is a strategic one intended to ease the way for possible encounters and dialogues that have been difficult to come by. I do not claim truth for the narratives and claims that follow, but rather that they can prove useful in opening up conceptual spaces for reflections, encounters, and dialogues for which many seem to yearn. In my experience, truth claims all too often have the effect of closing down conversations, of asserting arrival at a final account. What is needed in thinking about science and technology issues in today's world are, instead, courageous explorations into unfamiliar territories and far more tentative exchanges between peoples who rarely have occupied the same institutional locations.[9]

Overview. Chapter 1 sets up an issue about the relevance of postcolonial histories of science and technology to theories of knowledge and provides a preliminary map of the conceptual framework within which the book's discussions will occur. The next two chapters begin to reflect on accounts that have emerged from postcolonial science and technology studies, and especially their discussions of the causal relations between two familiar "marks of modernity": European expansion and the development of modern sciences in Europe. Chapter 4 suggests a way to conceptualize a point central not only to these postcolonial accounts, but also to the post-Kuhnian and feminist science and technology studies—namely, that all knowledge systems, including those of modern sciences, are local ones. It suggests a way to acknowledge and appreciate such arguments, while still retaining the conceptual ability to talk about the fact that many of modern sciences' claims, like some from the knowledge traditions of other cultures, are reliable under far more extensive historical, geographical, and cultural variations than are others: some knowledge claims are more powerful than others.

The next three chapters introduce feminist science and technology issues in both the post-Kuhnian and postcolonial streams and show how in significant respects they converge. The last four chapters pull together philosophic issues that have been raised in the earlier chapters. They begin to sketch out one way to retrieve and transform into tools useful in today's multicultural and postcolonial world some of the still immensely valuable elements of the European philosophic tradition.

ACKNOWLEDGMENTS

I am indebted to many institutions, colleagues, seminar participants, students, and lecture audiences for support and critical responses to this project over the last two decades. The opportunity in 1990 to teach a course at the University of Costa Rica and consult in Costa Rica, Nicaragua, Honduras, and Guatemala supported by the Women, Health, and Development project of the Pan American Health Organization provided the most compelling single stimulant for me to begin learning how to locate feminist science and technology issues on a multicultural and postcolonial map. I especially appreciated Yolanda Ingianna's efforts in making this possible. A fellowship in the Institute for Advanced Studies at the University of Delaware in 1989–90 permitted me to begin working up this book. The UCLA Center for the Study of Women and the Swiss Federal Institute of Technology Zurich (ETH) provided welcoming sites for further work on it in 1991. The United Nations Commission on Science and Technology for Development (UNCSTD), the United Nations Development Fund for Women (UNIFEM) and the United Nations Educational, Scientific, and Cultural Organization (UNESCO) provided opportunities for me to think about some of the issues of this book with development activists and scholars from around the world in several different contexts during 1994 and 1995.

Many colleagues, friends, and acquaintances have provided critical feedback on all or parts of the manuscript at various stages. Special thanks are due to Adele Clark, Anne Fausto-Sterling, Sarah Franklin, Valerie Hartouni, N. Katherine Hayles, Patti Lather, and Sharon Traweek. The editors of various journals and collections where earlier versions of some of the chapters have appeared have also been most helpful, and these publications are acknowledged at appropriate points in the manuscript.

I am grateful for ongoing conversations and exchanges through the years with friends and scholars working on similar issues, including Margaret Andersen, Arjuna DeZoysa, Cynthia Enloe, Susantha Goonatilake, Donna Haraway, Londa Schiebinger, Joni Seager, and Alison Wylie. I am indebted to Anne Fausto-Sterling and Sharon Traweek in this context also.

Finally, Dorian Harding-Morick, Emily Harding-Morick, and dear friends in Los Angeles and elsewhere have provided that nourishment for the soul without which thought falters.

1

A Role for Postcolonial Histories of Science in Theories of Knowledge? Conceptual Shifts

1. Are postcolonial histories of science relevant to theories of knowledge? Post–World War II science and technology studies. In his widely read study of scientific change, historian Thomas Kuhn asked "How could history of science fail to be a source of phenomena to which theories about knowledge may legitimately be asked to apply?"[1] However, it was not the older, "internalist" intellectual history of science to which he referred, but the new social histories that had begun to appear. The history of science that was to supply phenomena for philosophies, sociologies, and ethnographies of knowledge was no longer to be the familiar repository for "anecdote or chronology" gained from "the study of finished scientific achievements as these are recorded in the classics and, more recently, in the textbooks from which each new scientific generation learns to practice its trade." Instead, knowledge theorists—philosophers, sociologists, and others—were to draw on resources provided by the emerging "historiographic revolution in the study of science" that was learning how "to display the historical integrity of that science in its own time."[2] This project has taken on new meanings with the emergence of postcolonial science and technology studies.

Kuhn was just one of the researchers working in various disciplines in the decades following World War II whose treatment of modern sciences as historical, sociological, cultural, and political phenomena, ones no different in many respects from any other social institutions, their cultures, and practices, launched a revolution in science studies. During the same period, widespread public concern about science began to focus on how scientists should be educated, how class, race, gender, imperial and colonial interests and discourses had shaped scientific projects in the past and continued to do so in the second half of the twentieth century, and how destructive the consequences of scientific and technological change were in such phenomena as Hiroshima, the Viet Nam war, and Cold War politics, as well as for the natural environment and for workplaces. Stimulated by such concerns, political scientists, historians, philosophers, sociologists, ethnographers, and practicing scientists joined the historians in projects designed to display the historical integrity of modern sciences with the rest of the economic, political, and social relations of their eras. Sciences and their societies, it turned out, co-constructed each other.

Internalist scientific epistemologies and challenges to them. These studies, and their successors in subsequent decades, have raised fundamental challenges to the conventional epistemology of modern science. A central assumption of this older theory of scientific knowledge is that the success of modern science is insured by its internal features—experimental method or scientific method more generally, science's standards for maximizing objectivity and rationality, the use of mathematics to express nature's laws, the distinction between primary and secondary qualities in nature, or some other. Science is singular—there is one and only one science—and its components are harmoniously integrated by such internal features.[3] That internalist epistemology, as I shall refer to it, first began to take shape with the emergence of modern sciences five centuries ago, though it has been revised and refined many times since then. It is still the prevailing theory of scientific knowledge, disseminated in science texts and popular thought about the sciences and invoked by scientists when speaking to funding sources, Congress, or the general public. It also haunts the imaginations of some of the scholarly science studies still produced in history, sociology, and philosophy—though its overt supporters are becoming scarce in these fields. For those who actually have examined the mass of historical, sociological, and ethnographic evidence produced in more than a quarter-century of examination of the integrity of sciences with the economic, political, and social projects of their eras, the internalist epistemology has become a relic of modern western "folk belief." While it still is articulated with noble intent, no longer can the historical claims upon which its plausibility depends pass the kinds of empirical and theoretical tests to which other historical claims are expected to conform.

This older epistemology has assumed that the representations of nature's order that the sciences produce can be mirror-like reflections of a reality that is already out there and available for the reflecting, in Richard Rorty's phrase.[4] While increasing numbers of scholars no longer believe that the historical evidence lends plausibility to the idea that scientific descriptions and explanations could actually achieve such mirror-like perfection, many of them still think that *trying* to produce them is the best goal for the sciences. After all, many other ideals that are probably unachievable in practice—such as ending jaywalking or littering, or trying always to tell the truth—are nevertheless regarded as valuable goals of human behavior. It is worthwhile to act as if it were the case that as science progresses, its representations get closer and closer to such a singular and perfectly accurate reflection of nature's unique order, according to this view.

Therefore, when sciences function at their very best, their institutions, cultures, and practices, including scientific methods, will contribute nothing culturally distinctive to the representations of nature that appear in the results of research, this line of reasoning goes. One should try to produce scientific information in which one can find no culturally distinctive interests or discursive resources of the societies that have produced the research. Of course, society and the institutions, cultures, and practices of the sciences should be understood to provide the necessary conditions for sciences to do their work, but they should not influence the results of research in any culturally distinctive way. Any and all social values and interests that might initially get into the results of scientific research should be firmly weeded out as soon as possible through subsequent critical vigilance.

This older epistemology holds not only that a modern science with such goals alone is capable of providing such an account of nature's order, but also that it is capable in principle of detecting all of nature's laws. As the issue was put in the "unity of science" form of this epistemology, there is only one "nature," one truth about it, and one science; and such a science can in principle reveal the complete, unitary, and harmoniously integrated truth about a reality that is ordered in such a way as to be available for such an accounting.[5]

The post-Kuhnian social studies of science projects challenge this internalist epistemology that attributes all of the achievements of the sciences to nature's order plus the sciences' internal processes—especially scientific method, understood as sharply demarcated from any other methods of obtaining knowledge.[6] Yet they do not adopt the "externalist" position that society is entirely responsible for science's achievements and shortcomings—that science is simply a "dupe" of societies and their politics and that nature makes no contributions to scientific claims. Instead, they take what has come to be called, somewhat misleadingly, a constructionist approach, charting how sciences (plural) and their cultures co-evolve, each playing a major role in constituting the other, in bringing it into existence in the first place and maintaining it on a continuing basis, constrained in diverse respects by nature's order. The distinctive ways that cultures gain knowledge contribute to their being the kinds of cultures they are; and the distinctiveness of cultures contributes to the distinctively "local" patterns of their systematic knowledge and systematic ignorance. "Constructivism"— with its misleading suggestion that pre-existing, fully formed "societies" just make up ("construct") the representations of nature that they want regardless of how the world around them is ordered—is what this innovative thesis of post-Kuhnian science studies was called by those who interpreted it in such a misleading way, and this is the name that has stuck to it in popular thinking. Yet this approach would better be referred to as "co-constructivism," "co-evolutionism," or even "co-constitutionism" to emphasize how systematic knowledge-seeking is always just one element in any culture, society, or social formation in its local environment, shifting and transforming other elements—education systems, legal systems, economic relations, religious beliefs and practices, state projects (such as war-making), gender relations—as it, in turn, is transformed by them.

However, post-Kuhnian science studies are not the only such important stream of science and technology studies to emerge after World War II, or to arrive at such a co-constructivist account. Indeed, even this post-Kuhnian stream can sometimes appear parochial if one starts asking epistemological questions from outside the European-American intellectual traditions that certainly contained Kuhn's thinking and are still the prevailing ones in the science studies influenced by Kuhn's moment of disillusion with the internalist epistemologies. The internalist epistemology that Kuhn challenged is also in trouble in increasingly widespread discussions around the globe of "the epistemological crisis of the modern West." So it is fitting that this study will not position itself in the post-Kuhnian stream of science and technology studies, which from other, less eurocentric perspectives can be seen to identify some but not all of the important sources of modern West's epistemological crisis. Instead, it will start off from the borders between it and the postcolonial discussions that began to emerge during the same period. How do the conventional internalist epistemologies of modern science and also the post-Kuhnian co-constructivist accounts appear from such a standpoint, and how do the postcolonial accounts extend, enrich, and also conflict with the post-Kuhnian science studies?

Another voice: postcolonial science and technology studies. Postcolonial science and technology studies are only now gaining audiences in scholarly and more widespread public intellectual discussions in the United States and Europe. Though their origins go back to the 1940s and earlier, they now circulate through a recent outpouring of scholarly studies, conference reports, and science policy debates, including increasingly forceful statements by national and international agencies that have been reported in the media. Such events as the Columbian Quincentennial and the United Nations' Rio Earth Summit, both in 1992, also stimulated interest in this work. It is interesting to see how the kind of co-constructivist accounts developed in the northern studies have arrived from a different route and with somewhat different focuses in these accounts.

The chapters that follow will examine how the concerns of these studies have both converged with and diverged from those of post-Kuhnian science and technology studies. Included in such examinations will be important feminist components of both post-Kuhnian and post-colonial science studies: both contain distinctive feminist analyses that take a somewhat different angle from the analyses of their "prefeminist" colleagues. Moments of scientific and technological change are always sites of struggle over how the benefits and costs of change will be distributed. Gender projects interact with international, national, and more local ones in ways that shape the outcome of such struggles. In both post–World War II schools (post-Kuhnian and postcolonial) of science and technology studies, the feminist components are not only in alliance but also in conflict with their "fraternity." And the northern and southern feminist analyses are also both in alliance and in conflict with each other.

This book's concern will be to reflect on some of the epistemological lessons that can be learned from these complex, rich, and important convergences and divergences. How should philosophies of science be reshaped to account for modern sciences' history, achievements, limitations, and possible futures identified in these kinds of studies that share skepticism about the conventional internalist epistemologies of science? How could we best revise conventional ways of thinking about the knowledge traditions of other cultures in light of these revelations?

Though this study will not position itself inside the post-Kuhnian stream of science and technology studies, it will remain in continual alliance and dialogue with it. Continuing such a dialogue begun in preceding paragraphs, one could say that now, more than three decades after the appearance of Kuhn's account and the first studies for which his work (among others) provided the maps, postcolonial science and technology studies are beginning to create a second "historiographic revolution"—one stimulated by reflections on the integrity in their own times of sciences and technologies with their global contexts. The new postcolonial science studies bring fresh perspectives to charting the integrity of European sciences with Europe's global economic and political relations, to accounting for the history, culture, and practices of science and technology projects of non-European cultures, and to identifying relationships between European and non-European science and technology projects—especially since 1492.

How did the postcolonial science and technology studies originate? What is so innovative about them?

Postcolonial origins. Apparently the first clues to the possibility of postcolonial accounts of the dependency relations between the emergence of modern sciences and technologies in Europe and European expansion began to appear over fifty years ago. A West Indian historian looked at how the immense profits from Caribbean plantations had played such a large role in making possible industrialization in Europe. And an Indian historian began to examine how the British intentionally destroyed the Indian textile industry in order to create a market for the importation of British textiles.[7] Since scientific and technological knowledge, in Europe and in Europe's overseas targets of imperial control, were central to both historical moments (as we shall see in chapters to follow), it began to appear that scientific and technological growth in Europe were advanced through such processes in European expansion. Moreover, such expansion in turn appeared to be advanced by the development of modern sciences in Europe and the decline of "local knowledge traditions" in the Caribbean and in India. These prefigurings of the subsequently developed world systems theory contained important claims about how European expansion and the growth of modern sciences and technologies in Europe were causally linked.[8] In the 1950s and 1960s, challenges to the prevailing diffusionist model of scientific and technological growth, and to the presumed irrelevance of European imperialism to such processes, emerged in Europe, Australia, and other parts of the world.[9]

The new interest in causal relations between European expansion and the emergence of modern sciences in Europe has also helped bring into focus how scientifically and technologically advanced some other cultures have been relative to Europe's most advanced cultures—in some respects until well into the eighteenth century. Studies of contemporary non-European cultures' scientific and technological practices draw attention to some of their capacities and abilities that European sciences and technologies today lack—topics to be pursued further in the chapters that follow.[10] From the perspective of these studies, it is clear that modern, European sciences and technologies have not been the only highly accomplished ones in the past, that they have serious limitations today, and that there is little reason to think they will be the only highly accomplished ones in the future.

Indeed, the postcolonial studies have been able to bring into focus what a tragedy it would be should the human species arrive at one and only one universally valid scientific and technological tradition. The modern European epistemological dream of a perfectly coherent account of all of nature's regularities, one that perfectly corresponds to nature's order, is beginning to take on the character of a nightmare, as some of the post-Kuhnian studies as well as the postcolonial studies have grasped.[11] However, just what shape global sciences and technologies will take in the future cannot be clearly foreseen in any of these accounts since their future will be an outcome of struggles over their direction that surround and engage us even now.

Finally, the most recent critical analyses of the successes and failures of so-called Third World "development" have highlighted issues that have only been glimpsed otherwise. Development has been conceptualized in the North as the transfer of European models of industrialization—of European sciences and technologies—to the "underdeveloped" societies in the Third World. However, postcolonial development studies show that this kind of process has primarily de-developed the vast majority of the peoples who were supposed to benefit from such science and technology transfers. These accounts have drawn attention to how such development knowledge systems largely have been constituted by distinctively European and North American cultures' projects—including distinctively gendered ones—that insure a far more limited grasp of nature's order than conventionally has been imagined.[12]

Moreover, such studies show that we have appreciated modern sciences in part for the wrong reasons. It is not their ability to immunize their accounts of nature's order against all the cultural elements in their making and continued use that is responsible for their great successes, as has been assumed. Rather, it is the ability to neutralize some such cultural elements while fully exploiting others that has been responsible both for their successes and their failures. In those cases where they have been most successful, they have been able to maintain a crucial tension between maximally "global" and firmly local elements in the resources they bring to charting nature's order—though even the particular ways in which "global" elements are used are not culture free. From such a perspective, the "universally valid" versus "merely applied" science distinction appears to have no role to play in the history or philosophy of science since its useful work is done by the historically more accurate and philosophically more coherent accounts of fruitful tensions between the maximally global and the firmly local—a topic to be pursued in later chapters.

Postcolonial innovations. The account of the origins of postcolonial science and technology studies already points to important contributions they make to our understanding of the "integrity of sciences" with their distinctive global and local histories. One can conceptualize the innovations created by this kind of study in other useful ways.

For one, the postcolonial studies use as their evidence not the kind of European history and world history most of us learned but, instead, the now widely disseminated postcolonial, multicultural, global histories. In this context, the older form of history can be referred to as "isolationist history," for it recounted the histories of Europe, Africa, China, the Americas, and the societies of other parts of the world as largely separate and self-contained chronologies, more or less isolated from each other except for the one-way diffusion of the achievements of European societies to the others. In such history, Europe gets to appropriate ancient Greek achievements as the origins of modern European culture, and other cultures in today's world are presented as the latest stage in their own historical "tunnels" back through the centuries—that is, where any historical change at all is attributed to these other cultures.

In contrast, the postcolonial histories look at how cultures have been interacting with each other from the beginning of recorded human history. Cultures have exchanged shells, beads, seeds, cattle, manufactured goods, women, and scientific and technological ideas. Scientific and technological ideas have always been the easiest to transport from one culture to another and the quickest to travel through a receiving culture's networks. (They weigh very little, don't have to be fed, and don't run away!)[13]

This new kind of science studies differs from familiar kinds in yet other ways. Its subject—the "speaker" of these studies, the position from which they are organized—is not the "rational man" (European) from the perspective of whose experiences, interests, and desires isolationist histories were organized. Science in global history no longer is to be understood from the conventional eurocentric perspective that shaped earlier studies.[14] Instead, the conceptual framework of postcolonial science and technology studies is organized from the standpoint of other, non-European cultures and the great masses of the world's economically and politically most vulnerable people who live in them. Their scientific and technological needs and desires are not always those of elites in the North or in their own societies. On such "have nots" the development of modern sciences and technologies has had few beneficial and many detrimental effects. Thus, this postcolonial science theory organizes its concerns and conceptual frameworks from outside the familiar eurocentric ones and, in that sense, its "subject" or "author" is not the familiar enthusiastic European beneficiary of northern sciences and technologies. Such a strategy enables postcolonial theory to detect features of different cultures' scientific and technological thought and practices that are not visible from within the familiar western accounts of science. This new kind of account does not merely add new topics to conceptual frameworks that are themselves left unchanged. Instead, it forces transformations of them.

Postcolonial science studies differ from conventional and much of the more recent northern histories and philosophies of science in another way. The former has firmly rejected the role of disinterested observer of sciences in history—mere "handmaids" to the sciences, as John Locke put the point—that is often (though not invariably) the self-assigned role of northern science studies. Instead it engages with both discussions within the sciences about how best to represent nature's order and with science and technology policy. It joins scientists in critically examining which are the scientifically most defensible descriptions and explanations of nature's regularities; in this way it provides resources for more accurate and comprehensive scientific and technological thought. Such a project has implications for science policy, and the postcolonial studies are interested to propose which kinds of sciences will most advance both the growth of knowledge and the social welfare of the most vulnerable groups in their cultures. Thus, the postcolonial studies provide valuable resources for current public debates about sciences in a multicultural and global society—what kinds of sciences we have had and have now, what kinds we can get, and what kinds we could or should want. Since modern European sciences are desired, produced, and used around the world these days, their "world" is necessarily multicultural, and global international social relations as well as the more local ones that sustain them are also shaped by their effects. This is not to say that the post-Kuhnian accounts never enter such larger public debates; increasingly they are doing so. Rather, in contrast to the postcolonial accounts, their conceptual frameworks mostly have not been organized to serve such overtly "political" projects but, instead, the historical, sociological, and epistemological ones that they share with the postcolonial accounts.[15]

These are controversial issues, and we shall be pursuing them further in subsequent chapters. Here we can repeat Kuhn's 1960s question with a new inflection: how could the recent interactionist accounts of sciences and technologies in multicultural and global (and gendered) history fail to be a source of phenomena to which theories about knowledge may legitimately be asked to apply?

Before turning to examine in greater detail the claims of postcolonial science and technology studies, it will help to make an initial attempt to clarify central terms and claims in the post-positivist, co-constructivist, postcolonial, feminist conceptual framework within which this essay will explore its topics.

2. Conceptual shifts. a. Science. There are several reasons why readers may find it strange to use the term "science" to refer to beliefs and practices of non-European cultures, as will be the practice here. For one, such usage can appear to ignore important differences between achievements of European cultures during the last five centuries and what are often regarded as only traditional beliefs and practices of other cultures. Commitment to the difference in kind of the beliefs and practices advanced by European sciences is central to the self-conception of many people around the world as modern, enlightened, progressive, and guided in our beliefs and behaviors by the highest standards of objectivity and rationality. Thus, modern science has been conceptualized as contrasting with earlier European and non-European cultures' magic, witchcraft, pre-logical thought, superstitions or pseudosciences; with "folk explanations" or ethnosciences that are embedded in religious, anthropomorphic, and other only local belief systems; with merely technological achievements or merely speculative claims about the natural world; or with "precursors" to true sciences. Clearly, northerners have had a problem figuring out how to refer to the science and technology traditions of other cultures.

Perhaps there is something useful still to be found in such contrasts. However, whether or not this is so, they restrict the meanings and references of "science" in ways that are intellectually costly, as chapters that follow will explore. That is, one does not have to demonstrate that there is no longer anything at all useful in such contrasts in order to justify abandoning them or using them only very, very carefully in very limited contexts. One can instead point out that the costs outweigh the benefits of continuing to employ them. Such contrasts are often invoked only implicitly—for example, through locating the examination only of modern European sciences in a history or philosophy of science curriculum or course, while the study of "indigenous knowledge systems" must be pursued elsewhere, in the anthropology department. Through such practices do history, philosophy, and anthropology departments become complicit in eurocentric depictions of objectivity and rationality as at home in the West and "bias" and irrationality as at home with the rest of the world's cultures. These forms of institutional and social eurocentrism serve eurocentric interests even more effectively than do the overt references to witchcraft and superstitions, and so on, that educated middle classes are sometimes today taught to avoid in the name of tolerance. This study will use a more inclusive definition of science, one that encourages us to reexamine just when it is useful and when it is too costly to invoke a more restrictive definition. "Science" will be used to refer to any systematic attempt to produce knowledge about the natural world, just as "social science" refers to systematic attempts to produce knowledge about social worlds.[16]

However, this usage risks advancing eurocentrism on another front. Why should a cultural practice have value only if it can be squeezed into categories designed by Europeans to appreciate European institutions, their cultures and practices? Though the extension of this term to other cultures' science and technology traditions clears a space for countering eurocentrism and for advancing the growth of knowledge in one way, it can appear to promote eurocentrism and its forms of systematic ignorance in another way. Why not just leave "science" referring to European projects and critically appreciate in the terms used by other cultures their inquiries into nature's regularities and the best explanations of them? The Chinese, Arabic, and Andean indigenes did not refer to their systematic knowledge about the natural world as "science." (At this point northerners could recollect that neither did Europeans until fairly recently. "Natural philosophy" was the term used to refer to the accomplishments of Galileo, Kepler, Harvey, Boyle, and Newton until well into the nineteenth century. The term *scientist* was first used by William Whewell in 1840.)[17]

Attractive as such anti-eurocentric reasons are for continuing past practice, in my view they are not good enough to restrict the term in conventional ways only to sciences that have emerged in Europe. The contrast between science and superstition, precursors, and so forth, has been too useful a tool of eurocentric thinking to leave familiar usage a good choice for this study. Therefore, this study will stick with the more inclusive usage, though readers should keep in mind the way this enacts yet another piece of cultural imperialism that we will wish to move away from in other contexts. One should note here also that even this inclusive definition already is more restrictive than is common in Europe, where "science" is assumed to refer also to social sciences and even to the humanities. What is meant by "science" will be expanded and complexified in yet other ways in the chapters ahead—for example, as we come to see how knowledge about the natural world is intertwined with and even constituted by knowledge about the social world, and vice versa, in both the so-called natural and social sciences.

One of the conventional contrasts with "real science"—namely technology—deserves a few more words at this point. Does this study conflate science with technology when it consistently links science and technology studies, and when it refers to practices of other cultures as scientific that conventionally are regarded only as technological? The short answer is "no"; there are good reasons to link science and technology studies and to expand the referents of "science" to include systematic knowledge about the natural order produced in societies that do not have what is conventionally thought of as modern sciences. Most crucially, modern scientific knowledge is itself constituted through the technological practices of scientific research—a point mentioned in opening this chapter. The experimental method that is claimed to distinguish modern sciences, whether practiced in laboratories, field sites, or at computers, constitutes (co-constructs) scientific knowledge through technical interventions in nature. Thus, quite apart from how the information science produces is used, the very production of that information is technical. Moreover, such research technologies also often prefigure the applications and technologies for which such scientific information will be useful. Thus, scientific knowledge is inseparable from the technologies of its production, these have social and political preconditions and effects, and they provide blueprints for subsequent technological innovations.

On the other hand, there may well be reasons to discuss technological knowledge and its advancement without reference to advances in scientific knowledge, since it is clear that the former can occur without the latter. There was no new scientific knowledge necessary for the invention of the infant back carrier—the "baby backpack"—now so widely used in the West in the last quarter century. Historians of public health have argued that improved nutrition and public health practices, not advances in medical sciences, were largely responsible for the great decreases in mortality that occurred in late-nineteenth- and early-twentieth-century Europe and North America. Historians of science have begun to doubt that advances in scientific knowledge were responsible for as much of European industrialization prior to the late nineteenth century as is conventionally claimed. Thus, the growth of technological knowledge cannot be conceptualized as dependent upon the growth of scientific knowledge (and may even have its own epistemology), though the latter seems to be dependent on the former.[18]

Further on we will see how a number of other dualisms used to police the borders between "real sciences" and "not really sciences" are no longer useful or have far more limited usefulness than has been assumed.

b. Eurocentrism. What is eurocentrism? There is a widespread temptation to understand it as the overt or covert prejudices of individuals. Thus, eurocentrism is thought of as an expression of individuals' false beliefs and bad attitudes, just as are related cases of racism and sexism. However, it has long been demonstrated that the beliefs and behaviors of individuals are more adequately understood as the consequences, not the causes, of institutional, societal, and civilizational (or "philosophical") social structures and discursive assumptions. The prejudice account does not fit the empirical evidence, for the groups whose actions are most responsible for maintaining institutional, societal, and civilizational racism, sexism, and eurocentrism often are not at all prejudiced toward those who bear the costs of such discrimination. These administrators and managers of transnational corporations, of local, national, and international government agencies, of educational, research, and health care systems, and so on, often have good attitudes toward people of other races, genders, and cultures, and sometimes have relatively few false beliefs about them. Educated to understand how hurtful and unseemly it can be to express such false beliefs and bad attitudes, these groups practice "tolerance" instead.[19] However, such practices can often mask the far more powerful forms of institutional, social, and civilizational/philosophical eurocentrism that such groups are assigned to administer and manage.

Understood only as a set of beliefs and attitudes, eurocentrism is one of those socially powerful *incoherent* concepts, the usefulness of which is to be found in its incoherence.[20] When eurocentrism is understood as prejudice, the least-visibly eurocentric groups can, intentionally or not, most effectively advance eurocentric institutional, societal, and civilizational practices. To say this is not to attribute such eurocentric motivations inevitably to those who design and maintain eurocentric institutions, for while some have such covert intentions, others may not. Some may actively abhor eurocentrism and yet end up with beliefs and behaviors that advance it. Thus, good intentions and tolerant behaviors are not enough to guarantee that one is in fact supporting anti-eurocentric beliefs and practices. It is therefore useful for those who bear the costs of eurocentrism—and many peoples of European descent will correctly perceive themselves to be in this group, as we shall explore later on—to understand it as fundamentally a set of institutional, societal, and civilizational arrangements for distributing scarce economic, social, and political resources.

In pursuing the analyses of this study, it will help to keep in mind five forms that eurocentric and other such discriminatory beliefs and practices can take in order to distinguish which kind is the target of criticism in each analysis.[21] The first two forms, overt and covert eurocentric beliefs and practices, are intentionally enacted by individuals. The overt ones are openly expressed and practiced; covert ones are hidden by their perpetrators. For example, to dismiss openly the effectiveness of acupuncture as merely "folk belief" or of herbal pharmacologies as superstitions would be to practice overt eurocentrism. However, to dismiss them on the grounds that one's doctor does not recommend them, while believing that because they are non-European health practices they must be inferior, would be to practice covert eurocentrism.

Institutional eurocentrism occurs when medical associations do not admit trained acupuncturists or herbal pharmacologists, medical and pharmacology schools do not include training in such treatments, history of science and medicine journals will not publish accounts sympathetic to such practices, and history of science and medicine courses leave such traditions out of their accounts, perhaps sending students who request such information to the anthropology department or the ethnic studies program. Here, institutional practices have eurocentric effects upon both Europeans and non-Europeans. In such cases, perfectly well-intentioned individuals, even ones who understand the health value of such nonmodern European practices, advance institutional eurocentrism to the extent that they fail to challenge the conceptual frameworks that legitimate such discriminatory practices in these institutions.

Societal eurocentrism occurs when the kinds of beliefs evidenced by such institutional practices are in fact held by the larger culture that establishes and maintains the institutions mentioned in the preceding paragraph. The institutions' practices simply express widespread social assumptions. Thus, it would be a case of institutional and societal eurocentrism if virtually no one in the larger U.S. culture disagreed with such a low opinion of acupuncture and herbal pharmacologies, but only institutional eurocentrism if the medical associations, the medical and pharmacological schools, the journals, and history courses refused to recognize the value of these traditions in the face of obvious acceptance of such practices by otherwise "perfectly rational" and informed members of the larger society—the situation that is, one hopes, about to be avoided in these two cases in the United States as a few leading medical and health care institutions begin to legitimate these bodies of knowledge—these sciences—and their therapeutic technologies.

Civilizational or philosophic eurocentrism occurs when the beliefs and practices at issue are held by entire "civilizations" over large periods of history, not just by one of their subcultures. These are the most difficult to identify because they structure and give meaning to such apparently seamless expanses of history, common sense, and daily life that it is hard for members of such "civilizations" even to imagine taking a position that is outside them. "The scientific worldview," "the modern worldview," the Christian, Judaic, Islamic, Ancient Greek, or Chinese worldview provide examples of such widespread and long-term belief systems within which, in spite of great diversity of many beliefs and practices over time and between subcultures, distinctive shared or continuing beliefs and practices nevertheless can be identified. In such broad contexts, eurocentrism appears as an ethic, an ontology, and an epistemology; we can speak meaningfully of eurocentric ethical, ontological, and/or epistemological beliefs and practices. The assumption that modern science is trans- or a-cultural, and thus could not be multicultural or androcentric in any fundamental way, is just such a civilizational belief for most members of the educated middle classes in Europe, the Americas, and elsewhere around the globe. The "European diaspora" has greatly expanded the boundaries of eurocentrism.

As a civilizational practice, eurocentrism is a "discourse" in the rich, materialist sense that includes, but is not restricted merely to, ways of thinking or speaking.[22] Central among the presuppositions of eurocentric discourses are that peoples of European descent, their institutions, practices, and favored conceptual schemes, express the unique heights of human development. Moreover, peoples of European descent and their civilization are presumed to be fundamentally self-generated, owing little or nothing to the institutions, practices, conceptual schemes, or peoples of other parts of the world. These assumptions have organized in different ways in the last several hundred years—but especially since the eighteenth century—economic, political, historical, legal, geographical, archeological, sociological, linguistic, anthropological, psychological, pedagogical, literary, art historical, philosophical, biological, medical, and technological institutions and their practices.[23]

What is most startling, and disturbing, from such a perspective of institutional, societal, and civilizational eurocentrism is to realize that even individuals with the highest moral intentions, and with the most up-to-date, state-of-the-art, well-informed, rational standards according to the prevailing institutions and their larger cultures, can still be actively advancing institutional, societal, and philosophic eurocentrism. The prevailing institutional and cultural standards turn out themselves to be significant obstacles to identifying the eurocentrism of institutional, societal, and civilizational beliefs and practices.[24] It is these more extensive, harder to detect, forms of eurocentrism with which this study is concerned.

c. *Postcolonial.* This brings us to the term *postcolonial.* For most beneficiaries of colonialism, the colonial conceptual framework functioned at institutional, societal, and perhaps even civilizational levels. While such beneficiaries could always imagine postcolonialism as a possible future nightmarish state of affairs against which they had to defend themselves, there was no hint in the thinking of most of them that anticolonial institutions and their underlying assumptions could be reasonable and/or desirable ones for people like themselves.

The term has had many other referents and meanings. There is also its temporal reference to the period of time that began in the 1960s, marked by the end of formal European colonialism, that will persist indefinitely far into the future—the postcolonial era. There is the condition of being no longer formally a colony or member of such a colony of one of the European states; new nation-states and their citizens are postcolonial. However, a variety of concerns and stances called postcolonial also appear in discussions about what are the most accurate and desirable understandings of colonialism, what are and should be the present relations between ex-colonials and their former colonizers, and what should be their future relations with each other. For example, there is the postcolonialism that is still imagined to lie only in the future by those who find the contemporary development policies of the international agencies and northern nations merely "colonialism by other means." There is the postcolonialism that is a return to or revival of pre-colonized voices, institutions, cultures and practices by the formerly colonized. There is the postcolonialism that appeared within colonialism as a critical counterdiscourse by the colonized, and the different postcolonialisms that appeared within colonialism by those with the privileges of the colonizers—the discourses of the critics, protesters, dissidents. These postcolonialisms, along with "the end of colonialism" mentioned above that is always imagined as a possible future feared by the colonizers, were features of colonialism itself.

The critical counterdiscourse by the colonized can appear either as an oppositional discourse, by those who, say, actively work to overthrow the rule of the colonized, or as a more ambivalent, complicitous discourse by those who criticize the evils of colonialism even as they also extol its virtues and its necessity. There is the postcolonial critique of those in hybrid conditions at the borders between the colonizers and the colonized—educated Indians in London, or French visitors to the eighteenth-century American South. Today many students, scholars, and citizens in Europe and the United States provide such postcolonial critiques from these borderland locations.[25] Another kind of postcolonialism appears in the critique by those settler colonials who themselves were also colonizers of the indigenes—for example, in the successors to "commonwealth literature" from Australians, Canadians, and British South Africans, but also present in the Spanish, Portuguese and French Americas, and other such places around the globe. Moreover, there are forms of postcolonialism linked to postmodernism and/or to postorientalism, and forms that are not.[26]

These different historical positionings with respect to colonialism create different concerns in the postcolonial science and technology studies also, and it is not always easy to sort them out without more detailed knowledge of the histories of their development. However, there are three guidelines one can extract from even this mere glimpse of the diversity of meanings and referents for the term and the complexity of the conceptual and political terrains on which it has been found useful. One is the obvious observation that postcolonialism is not monolithic, and the diversity of its concerns and stances provides valuable resources for thinking about the social and historical contexts in which scientific and technological changes occur. What looks like a reasonable claim or practice in one context can be problematic in another. A second is that we might do best to follow anthropologist David Hess's lead and think of "the postcolonial" as a kind of critical discursive space opened up both within and after the end of formal colonialism, where diverse positionings, discussions and other practices can occur.[27] We can employ the category of the postcolonial strategically as a kind of instrument or method of detecting phenomena that otherwise are occluded.

Third, it can be useful to speak of "decolonization" and "decolonizing" as a distinctive political and intellectual tendency within postcolonial spaces and their diverse discussions. Such terms draw attention to the necessity of active intervention in still prevailing and powerful discourses, their institutions, and practices, in order to end the forms of colonialism and neocolonialism that still structure most people's lives in the North and everywhere else around the globe. Such terms counter the tendency to think of the postcolonial as a kind of state of grace from which we lucky people get to benefit without exerting any political or intellectual effort. This book is most interested in science studies that aim to decolonize thinking about sciences, nature, and history and, thus, the ongoing practices of sciences and their technologies in the North and the South.

d. *Postcolonial and feminist standpoints.* This study will operate from a postcolonial and feminist *standpoint.*[28] What does this mean? For one, as the discussions above reveal, these adjectives mark critical and theoretical positions or possibilities. *Postcolonial* is not a geographical, national, or racial category, nor is *feminist* a women's identity, let alone a biological category. *Postcolonial* and *feminist* are not identities in the sense of pre-existing natural or social roles into which one is born or which one otherwise unreflectively acquires. However, if one conceptualizes identities in a different way, as commitments to chosen political struggles or to distinctive visions of the future, as has been widely argued for example about such identities as "black" and as "chicano," then "postcolonial" and "feminist" could usefully be thought of as identities. No one is born black, chicana, postcolonial, or feminist, in this way of thinking. Nor can anyone simply exert one's will power and choose one or more of these as a legitimate identity. Nevertheless, these are chosen identities.[29]

Less problematically, postcolonialism and feminism can usefully be thought of as thinking spaces that have been opened up by changes in social relations and in ways of thinking about them—by changes in "discourses." Within such spaces, new kinds of questions can be asked and new kinds of possible futures can be articulated and debated. Starting thought from the lives of those people upon whose exploitation the legitimacy of the dominant system depends can bring into focus questions and issues that were not visible, "important," or legitimate within the dominant institutions, their conceptual frameworks, cultures, and practices.[30] Postcolonialisms and feminisms articulate such standpoints. Important postcolonial analyses have been produced by Egyptian, English, Afro-Caribbean, Pakistani, Brazilian, French, U.S., and Sri Lankan citizens, among others, of many "races," ethnicities, and mother tongues. Important feminist accounts have been produced by men.[31] Thus, such standpoints are critically and theoretically constructed discursive positions, not merely perspectives or views that flow from their authors unwittingly because of their biology or location in geographical or other such social relations.

This study is only *a* standpoint, not *the* postcolonial and/or feminist standpoint. The standpoint of this book is itself historically locatable in just the way that are the cultural histories, their practices and meanings, that it examines. (Though it is not thereby epistemologically relativist, as paragraphs below and later chapters will show.)

Such an epistemological stance is positioned against the "positivist," internalist epistemological ones that, explicitly or implicitly, largely continue to frame systematic knowledge-seeking projects about nature and society. However, it is also not an ethnography of the scientific and technological cultures of other societies, though it will draw on such ethnographies for evidence for its claims. It does not seek to show the "rationality" of other scientific and technological cultures from the perspective of their inhabitants, as ethnographies so often have done. It does not speak *for* others, for peoples from non-European cultures, though it is informed by their accounts. It does not "study down," sympathetically describing for metropolitan audiences the beliefs and practices of peoples located at their peripheries, though it uses such accounts as evidence for its claims. These kinds of projects, with which this study might be conflated, can be worthy ones (or not), but they are not the project of this study.

Instead, the standpoint from which this work is organized is one shared with other projects interested in countering eurocentric and androcentric science and technology policies and their effects. It intends to bring into clearer focus new questions but, like all standpoints, is not usually able to answer them in any final way. Instead, I hope that it generates wider discussion of crucial issues that were either invisible, considered unimportant, or delegitimated. I further hope that it can advance the growth of knowledge by making visible aspects of nature, sciences, history, and present-day social relations that are hard or impossible to detect from within the ways of thinking familiar in the dominant European and North American institutions, their cultures, and practices.

e. Strong objectivity: In opposition to epistemological relativism. Does using postcolonial and feminist standpoints necessarily decrease the objectivity of the arguments that follow? Does it commit this study to a relativist position? For many readers, these may seem like the necessary consequence of abandoning the familiar internalist epistemology and framing this work within postcolonialisms and feminisms. However, they would be wrong to draw such a conclusion. Instead, this study is committed to strengthening the objectivity of understandings of modern sciences and technologies, of the sciences and technologies of other cultures, and of historical and possible future relations between them. Identifying eurocentric and androcentric elements in the conceptual frameworks used to think about scientific and technological change, and in sciences and technologies themselves, expands our knowledge of nature, sciences, and social relations. "Starting thought from marginalized lives," as standpoint epistemologies recommend, thus provides more rigorous, more competent standards for maximizing objectivity.[32]

Such a program for stronger standards for objectivity draws attention to the sociological or historical relativism of all assumptions and knowledge claims—even the most abstract and apparently transcultural ones of modern European sciences. Assumptions and claims always originate from the projects of some particular culture at a determinate historical moment, and they continue to prove useful, or not, to these and other later cultures. However, the strong objectivity program rejects the epistemological or judgmental relativism that assumes that because all such assumptions and claims have local, historical components, there is no rational, defensible way to evaluate them. It rejects the idea that all claims are equally valid, that all cultures' science and technology projects are equally defensible, for any and all purposes. It rejects the assumption that if one recognizes the social, historical relativism of knowledge claims, one is forced to epistemological, judgmental relativism.

One problem with such a justification for epistemological relativism is that every epistemological position—internalism's absolutism no less than the historical relativism, as well as standpoint approaches—assumes that different cultures, peoples, or eras hold different beliefs about nature and social relations. Recognition of historical or sociological relativism does not force the absolutists to a relativist position; they hold that, nevertheless, there is one and only one defensible knowledge system, namely that of modern Europe. The standpoint approach, with its strong objectivity program simply disagrees: different cultures' knowledge systems have different resources and limitations for producing knowledge; they are not all "equal," but there is no single possible perfect one, either.

Thus, this study's third position conflicts with the epistemological relativist's about the logical consequences of historical/sociological relativism. Not all proposed standards for knowledge are equally good —indeed, some are not only inadequate, but dangerous to their believers' lives. One can easily be killed by poisonous foods, wild animals, excessive availability of handguns, drunk drivers, toxic environments, dangerous and faulty technologies, and cigarettes, for example, if one does not carefully evaluate the standards that friends, strangers, and diverse institutions use to sort knowledge claims into the reliable and the unreliable. However, there also is not just one adequate standard for knowledge, but different ones for different purposes. Some standards for adequate knowledge claims produce a rapid growth in knowledge about causes of cancer to be found within individuals and their lifestyles, and they also produce systematic ignorance about environmental causes of cancer for which military, governmental, and transnational economies are responsible. Other standards for adequate knowledge claims could produce different patterns of knowledge and ignorance about the causes of cancer.[33]

Hence, the postcolonial and feminist positioning of this study is precisely what advances its commitment to stronger standards of objectivity and against epistemological relativism.

f. After "realism versus constructivism." It should not need to be said, but probably does, that the choice between absolutist forms of realism and constructivism offers only inadequate options from the perspective of this study. Of course "there is a world out there," "reality exists," and successful, useful sciences and technologies, modern or not, have to be good at grasping a great deal about the realities of the parts of the world with which they interact. However, scientific and technological ideas and principles, too, bear distinctive marks of their social production and uses, though, as indicated earlier, "co-constructed" is a better way than "constructed" to think about the relationship between cultures and their knowledge systems. The old duality of "realism versus constructivism," like the others mentioned, has become an obstacle to our understanding of nature's order and the resources and limitations of current practices and proposals for developing knowledge about it. We can retain the best of both realist and constructivist understandings of the relations between our social worlds, our representations, and the realities our representations are intended to represent by thinking of co-evolving, or co-constructing, cultures and their knowledge projects.

g. *After the universality ideal.* Similarly, we can ask what is gained and what is lost by conceptualizing scientific and technological claims as universal versus only as merely local ones. The argument here will be that cultures are not only "prison houses" for the growth of scientific and technological knowledge, as they have usually been conceptualized. They are also "toolboxes" for such projects. Cultures generate scientific and technological projects to serve distinctively local interests and needs in the first place. Moreover, the diversity of the cultural resources that they bring to such projects enables humanity ever to see yet more aspects of nature's order. Cultures' distinctive ways of organizing the production of knowledge produce distinctive repositories of knowledge and method—through different kinds of laboratory or field experiments, through the "voyages of discovery," through farmers', travelers', mothers', and cooks' daily practices over time. The limit of such resources can never be reached as long as cultures continue to change over time and new ones emerge in the diasporas and interstices of older cultures. Thus, we would do better to think of scientific and technological claims as located on a continuum where "global" occupies one pole, "local" the other, and "universal" disappears as no longer useful.

Does it need mentioning that, of course, nature's "law of gravity" will have its effects on us whether we are Chilean or British, Catholic or Moslem, masculine or feminine, and whether or not we are aware of or believe it? What is at issue in the universal science disputes is not such phenomena, but whether there is one and only one best way for all purposes, now and in the unforeseeable future, to represent such an aspect of nature's order.[34] These issues, too, will be explored further below.

h. *Robust reflexivity.* And what is the epistemological status of this account? It, too, must be understood as a local knowledge system, developed from a determinate location in contemporary social relations and available discursive resources. It must acknowledge for itself the same critical standards it proposes for everyone else's knowledge systems. It will have to meet "robustly reflexive" standards—the very conditions it claims most advance the growth of knowledge in the scientific and technological cultures it examines. It cannot exempt itself from the conditions it identifies in the best historical examples of the growth of knowledge, as is characteristic of the "positivist," internalist epistemology as well as the uncritical, epistemologically relativist positions to which it objects. It must understand itself as shaped by social, historical conditions, just as it does the epistemologies and sciences it examines. And yet it must nevertheless be able to provide plausible evidence for its claims. This is what is required by a "robust reflexivity"—another issue to be explored below.

i. Strategic categories and concepts. Finally, it follows from the considerations raised above that the necessarily simplifying categories to be used in this account should be understood not as claims to "name reality" in some authoritative, perfect manner, but rather as ways of gaining a fresh perspective that can bring to our attention aspects of scientific and technological cultures and the worlds within which they exist that would otherwise be hard to detect. Thus, these categories are strategic rather than ontological. For example, the question here is not "are sciences (and our accounts of science) really realist or really constructivist?" but, rather, what can we learn about nature's order and about sciences by employing such frameworks, and what kinds of systematic ignorance are created by imagining there to be only two such discrete categories into which we are supposed to stuff all claims to empirically and theoretically adequate beliefs and practices? The categories and concepts found most useful for the project of this study at this time may well not be the best ones for other projects or in other contexts.

3. Conclusion: Political stakes. These are some of the main conceptual landmarks that have helped to map the space within which the concerns of this study will be pursued. Such maps, and the concerns that they make it possible to pursue, have more than scholarly consequences.

As the old saying goes, knowledge and power are intimately linked. That is nowhere more obvious than in the global political economy today. It is commonplace to note that since World War II, the base of this economy has shifted from heavy industry to information technologies and service industries. Thus, scientific innovation has moved even more firmly to the base of the contemporary economy. Whoever already owns "nature" and has access to it, whoever has the capital and knowledge to decide just how they can best access nature's resources and how such resources will be used—these are the peoples to whom the benefits of contemporary scientific and technological change largely will accrue. The majority of the world's peoples, in the North and the South, and especially women in every culture around the globe, have few of these resources. They do not own parts of nature; they do not have the resources to access its energies and powers; and they are systematically denied access to the knowledge of how to gain access to such parts of nature or technical resources. Since moments of scientific and techno- logical change are always sites of political struggle over who shall get the benefits and who bear the costs of such changes—and over who gets to make such decisions—the majority of the world's peoples will lose such struggles. Under such circumstances, it is difficult to understand how "more science and technology" of the kinds favored by the world's "haves" can fail to further enlarge the gap between themselves and the "have nots." This is a disturbing recognition for all of us who thought that more science and technology could advance *human* welfare and *social* progress, not the welfare and progress predominantly of the economically and politically most well-off at the expense of the welfare and progress of the vast majority of the globe's populations that are already the most economically and politically vulnerable.

Thus, scientific and technological changes do play central roles in advancing or blocking processes of global democratization. And the dreadful scenario just sketched out appears to be worsened as "mo- dernity," with its distinctive scientific and technological worldview, is further disseminated around the world as part of the advance of western economic forms and the spread of western forms of democratization. The scientific worldview travels with both of these and is strengthened by them, as has always been the case, according to the postcolonial accounts. The internalist theories of scientific knowledge that are the target of criticism by so many schools of post–World War II science studies are inadequate as guides for historians and philosophers to how science works. But, even worse, they may well be dangerous to human life, the environment, and maximally democratic social relations. The stakes in working to develop empirically, theoretically, and politically more useful science and technology theories are higher than one might at first have imagined.

Fortunately, it is not only the economically and politically most vul- nerable who can recognize the limitations of the conventional concep- tual frameworks. Increasingly, powerful groups within the North are coming to see that their pride in northern cultures, and their legitimacy as rational participants in local, national, and international institu- tions—not to mention their life and health—depend upon reworking such frameworks. It is a propitious moment to avail ourselves of the resources provided by postcolonial and feminist standpoints in order to try to make a difference to the outcome of this emerging reevaluation. In an important sense, the postcolonial and feminist standpoints are also firmly *within* conventional dynamics of northern sciences committed to critical self-evaluation.

Let us turn to pursue the postcolonial narrative, as promised earlier.

2

Postcolonial Science and Technology Studies:
A Space for New Questions

1. After "the tunnel of time": The emergence of interactionist global histories.
Since the beginning of the end of formal European colonial rule in the
1960s, a new kind of global history has emerged—one that has charted
the continual encounters and exchanges between cultures from the
beginnings of human history through the present day.[1] In the older
accounts the history of Europeans and the European diaspora overseas
was represented as largely isolated from the histories of other cultures
around the globe. Historical time for peoples of European descent was
represented as a tunnel stretching from the present back to the Garden
of Eden—"the tunnel of time." Everything important that ever happened to humans happened in that tunnel. In ancient times, "the Bible
lands" were included in the tunnel, but subsequently they disappeared
from European history. Other peoples had their own tunnels—Chinese,
Indians, Africans. But little or nothing of importance in human history
happened in those tunnels. Indeed, it was not always clear in these older
accounts that other cultures had a history at all, since they were represented as largely unchanging, timeless, in their social relations.[2]

In these accounts, which prevailed in textbooks until the last decade
or so, the interactions of modern western cultures with those in other
parts of the world were presented as mainly limited to the diffusion of
western accomplishments to other people mentioned there. As geographer J. M. Blaut pointed out, "the tunnel of time" and its diffusionist assumptions justified European colonialism with the following reasoning:

> The most important tenet of diffusionism is the theory of "the autono-
> mous rise of Europe," and sometimes (rather more grandly) the idea of
> "the European Miracle." It is the idea that Europe was more advanced and
> more progressive than all other regions prior to 1492, prior, that is, to the
> beginning of the period of colonialism, the period in which Europe and

non-Europe came into intense interaction. If one believes this to be the case—and most modern scholars seem to believe it to be the case—then it must follow that the economic and social modernization of Europe is fundamentally a result of Europe's *internal* qualities, not of interaction with the societies of Africa, Asia, and America after 1492. Therefore: the main building blocks of modernity must be European. Therefore: colonialism cannot have been really important for Europe's modernization. Therefore: colonialism must mean, for the Africans, Asians, and Americans, not spoilation and cultural destruction but, rather, the receipt-by-diffusion of European civilization: modernization.[3]

In contrast, the postcolonial accounts have made visible the continual interactions and exchanges between cultures around the globe, and the effects of such interactions on how cultures emerge, are transformed, and decline. Scientific and technological changes and exchanges are part of such histories, though they have been slower to come into clear focus.

This kind of postcolonial political and social history has already been widely adopted in U.S. K–12 history curricula.[4] While accounts of the interchanges between different cultures' science and technology projects have not yet reached such wide circulation, they can already be glimpsed in popular culture in media accounts of North versus South debates at the U.N.'s 1992 Rio Earth Summit, as well as in some of the museum exhibits, scholarly books, and media presentations emerging at the time of the 1992 Columbian Quincentennial commemorations. For example, at a Smithsonian exhibit entitled "1492" that consisted of separate rooms of artifacts representing the achievements of different cultures around the world by that momentous year, the European room contained scientific instruments that Europeans would bring to their world travels—beautifully crafted compasses, clocks, astrolabes, and models of the heavens. It also contained European-made maps of Europe, the Mediterranean, the coasts of Africa and Asia, though not yet of the Americas, of course. Visitors could not help but reflect on how the "Voyages of Discovery" would soon enable European scientific and technological knowledge to multiply at exponential rates, and on how archaic the beautiful maps soon would become. Moreover, one can note that Blaut's books (cited above) were published just at this time. Since modernization is always defined in terms of a distinctive worldview in which the preconditions and achievements of modern sciences and technologies play a central role, the diffusionist debates are necessarily also about modern sciences and technologies.

Postcolonial science and technology studies have origins separate from the innovative northern science and technology studies that began appearing during the same period, and for which the publication in 1962 of Thomas Kuhn's *The Structure of Scientific Revolutions* can serve as a convenient starting point (even though a number of other important philosophers, sociologists, and historians also played important roles in its takeoff.)[5] However, these northern, post-Kuhnian studies make the postcolonial accounts far more plausible. The previous chapter outlined some of the central conceptual shifts created by the post-Kuhnian accounts. It also identified three central focuses of today's postcolonial science and technology studies: the relationship between scientific and technological change and projects of European-American empire, anti-eurocentric accounts of other cultures' scientific and technological traditions, and the implications of the now obvious failures of the North's attempts to increase the standard of living in the South by transferring northern sciences and technologies to the South—the failures of "development." The sections that follow will show how the plausibility of the accounts emerging from these focuses has been greatly enhanced by post-Kuhnian re-examination of early modern European history.

The sections that follow examine first the rise in skepticism about the "European miracle," "the Dark Ages," and "the scientific revolution"—three concepts central to the internalist, isolationist histories of European and of other cultures' sciences and technologies. In the next two sections, new questions posed by the older Marxian externalist histories of science and technology, and then in the early forms of Marxian world systems theory, are examined. The fourth section looks at the early comparative ethnosciences accounts of contributions other scientific traditions made to the development of European sciences. The last section identifies the increasing implication of European sciences and technologies in the criticisms of post–World War II, Third World so-called development—criticisms that are themselves informed by some of the other early studies. In these kinds of accounts can be found reasons for the increased plausibility of today's postcolonial science and technology studies.[6]

However, this is not an attempt to map out these discussions in a comprehensive way, let alone to undertake the inappropriate task of authoritatively settling any of the many disputes and ambivalences rippling across the surfaces and through the depths of these postcolonial projects. Rather, this chapter (and the next) attempt to provide a series of snapshots highlighting key questions about the separate stream, internalist, diffusionist, histories of science and technology that can now be raised thanks to postcolonial science and technology studies, at significant points supported by post-Kuhnian studies emerging during the same period. These questions generate challenges to the historians, sociologists, ethnographers, and philosophers whose theories about knowledge are supposed to be able to account for the actual history of what everyone counts as modernity's model of knowledge—scientific and technological knowledge.

The five tributaries creating today's postcolonial studies that will be explored are not completely discrete; some authors and themes play a role in more than one of them. Each area of study has had concerns and accomplishments beyond those that will be described for the purposes here; thus, none are as narrowly focussed as this account of them may suggest. Moreover, though many of their concerns and analyses do converge, I do not mean to suggest that there are no political or scholarly disagreements between them, for there are many such.[7]

Finally, it is useful to recollect that "northern" and "postcolonial" science studies here refer to distinctive sets of problematics, themes, and standpoints—to discursive spaces—not to the ethnicity, nationality, or place of residence of the authors of these accounts. The postcolonial authors come from the United States, Europe, Australia, Africa, Latin America, Asia, and every other part of the world where European imperialism and, now, the emergence of postcoloniality have left their marks.[8]

With these reminders in place, let us turn to look at the origins of central themes in today's postcolonial science studies.

2. Problems with diffusionist, internalist histories. a. The European miracle, the Dark Ages, and the scientific revolution. Three key concepts in these internalist accounts had to be invented to maintain the internalist story: the European miracle, the Dark Ages and the scientific revolution. Before the early modern era in Europe, leadership in scientific and technological knowledge had shifted from one cultural center to another around the world—from China, to India, to Islamic cultures. Then it moved to Europe. However, as long as the "birth of modern science" was conceptualized as a sudden and momentous event, it was difficult to find any sufficiently powerful causes of it in the so-called Dark Ages. Where in medieval Europe could be found the unique latent potential for the great scientific, technological, and accompanying economic and cultural leaps that Europe subsequently took? Indeed, once historians began looking more carefully at other cultures, it would become clear that Europe lagged behind some of the other cultural centers in some economic, political, technological, and scientific respects and was no more than on a par with conditions in China and India in other respects.[9] Subsequently, the historians asked, looking at the state of sciences and technologies in early modern Europe, why did modern sciences and technologies arise in Europe rather than in one of the equally or perhaps even better-positioned earlier centers of scientific and technological innovation? All three concepts central to the internalist accounts began to attract skeptical attention.

The "European Miracle." The conventional strategy in answering this question was to seek "internal" features of European cultures and of their scientific and technological traditions to which could be attributed the "European miracle" and its development of modern sciences and technologies. Internal features of other cultures and their scientific and technological traditions that purportedly could account for the inherent unsuitability for developing modern sciences and technologies and the subsequent decline of these cultures were also identified. Blaut describes this view this way:

> Most European historians believe in some form of the theory of "the European miracle." This is the argument that Europe forged ahead of all other civilizations far back in history—in prehistoric or ancient or medieval times—and that this internally generated historical superiority or priority explains world history and geography after 1492: the modernization of Europe, the rise of capitalism, the conquest of the world. Most historians do not see anything miraculous in this process, but the phrase "the European miracle" became in the 1980s a very popular label for the whole family of theories about the supposedly unique rise of Europe before 1492.[10]

These theories persisted even though there was no agreement about the causes of the miracle:

> The historians do not agree among themselves on the question *why* the miracle occurred: why Europe forged ahead in this perhaps miraculous way. Is it because Europeans are genetically superior? are culturally superior? live in a superior environment? Is it because one special, wonderful thing happened in Europe, or happened to Europeans at a special moment in history, giving Europeans a decisive advantage over other societies?[11]

Why it occurred is just one topic on which historians disagree, for they also cannot reach consensus even about when it began. With the hindsight provided by several decades of postcolonial history that was emerging as the miracle accounts declined, one can see that there are good reasons for these disagreements. There was no such miracle, so of course it was difficult to identify its dates or causes. That is, early modern scientific and technological knowledge had strong roots in medieval European and Islamic cultures, irrational though these latter appeared to later historians. Moreover, it developed rapidly due to the need for scientific and technological information to speed along the Voyages of Discovery, the expansion of urban centers, and industrialization within Europe. (Its development was assisted, as we shall shortly see, by the resources European sciences and technologies gained from borrowings from the scientific and technological traditions of the cultures that expanding Europe encountered.)

The internalist accounts also distorted northern understandings of other cultures that had already developed advanced mathematical, theoretical, and technological traditions. In these cases the internalist historians argued that such cultures were capable of, or interested in, only the purely speculative or merely technological. Conversely, they claimed, these other cultures were unable to extract their scientific and technological concerns from their religious or cultural superstitions, prejudices, and projects—for example, from astrology, alchemy, anthropomorphism, or magical thinking. Some attributed to non-European peoples the capacity only for pre-logical thought patterns.[12] It was also claimed that these cultures could not develop the appropriate political climate for self-critical thought that was so crucial for the development of modern sciences.[13] The postcolonials point accusingly back at the larger conceptual frameworks of modern European cultures—their conceptual eurocentrism, androcentrism, service to class interests, and faith in Enlightenment assumptions—that consistently assume that for one reason or another, other cultures just did not have "the right stuff" to think up and develop modern sciences. But Europeans and European culture did, according to the internalist accounts. European mental abilities and talents, character, inquisitive and critical spirit, and political and cultural histories, institutions, and practices were just what was needed for the scientific and technological takeoff.

These internalist accounts also invoked the legacy of the classical Greek culture as an explanation of the European miracle. In this Greek culture what could reasonably be called scientific methods of observing and explaining the natural world initially appeared. The first stirrings of the idea of an experimental method occur in Aristotle's observations of the development of chicken fetuses. After the decline of classical Greek and then Roman culture, this Greek legacy was stored for centuries in Alexandria and was finally retrieved by its rightful heirs: the Europeans of the early Renaissance. Europe simply forgot where it left its rich scientific origins for more than a millennium, according to the standard account, but fortunately this legacy lay untouched in Egypt, awaiting recovery in the European Renaissance and Enlightenment. Thus, it was implied that the flourishing Islamic cultures in which the Greek legacy was saved those many centuries made no significant contributions to Europe's sciences. They just let the Greek legacy collect dust, preserved in its original form, until Europe was ready to reclaim it. Such assumptions depend also on the concept of medieval Europe's "Dark Ages."

The "Dark Ages": Just who was "dark"? Conveniently overlooked in these accounts was the way such a designation for premodern Europe was more a reflection of subsequent scholarly ignorance about this period than of the accomplishments of the period itself. The more premodern Europe began to be examined with the tools of modern history, the more accomplished and permeated with non-European cultures it appeared, and the more ignorant appeared those who saw it only as a dark space in the lineage of otherwise bright European history. Moreover, the miracle story ignores the fact that Europe did not even come into existence as a politically and culturally recognizable area until long after classical Greek society had disappeared—until Charlemagne's creation of the Holy Roman Empire, or, as other historians have argued, until the fifteenth to seventeenth centuries with the emergence of the early modern period itself. Aristotle's society was part of a Mediterranean culture that included North Africa and the eastern Mediterranean, and excluded most of what counts as Europe today. Another issue was that the Greek inheritance could also legitimately be claimed by many other cultures where the spread of Islam had diffused the Greek legacy to the Middle Eastern, Indian, other Asian, and many African cultures. Aristotle is a founding figure in several non-European cultural histories.

Thus, the purportedly purely Greek legacy that Europeans retrieved from the Alexandrian libraries had been created not only by Greeks some millennium and a half earlier, but also by all the intervening cultures in Africa, Asia, and the Mediterranean that had revised and enriched it. Once historians began exploring seriously the positive and constitutive effects on the growth of modern science that Egyptian mystical elements and indigenous European alchemy and astrology had provided, it became clear that modern notions of rationality could no longer be said to have played the lead role in birthing modern sciences.[14]

Was the scientific revolution really a revolution? As the evidence began to disappear that would support such notions as the European Dark Ages and subsequent miraculous flourishing, so too did it disappear for the revolutionary status of early modern scientific and technological knowledge. Conceptualizing "the scientific revolution" as a distinctive event began to appear to be more an artifact of subsequent eurocentric histories than of the actual historical record. The emergence of modern sciences and technologies in Europe was a slow and gradual process. Moreover, medieval Europeans, like their early modern descendants, were in continual contact and exchange with other cultures in India, China, and other parts of the world. Here let us turn to follow this thread as it appears in the Marxian externalist accounts of science and technology and in "world systems theory," both of which had beginnings in the 1930s and 1940s.

b. Externalist accounts. A second important step toward an alternative to the eurocentric accounts were the so-called "externalist" ones produced by European science scholars working in the Marxian tradition. Such accounts also have their limitations, but they did draw attention to the positive effects that cultural events and processes have had on the growth of modern sciences and technologies. It was not some distinctive collection of internal, epistemic characteristics of European minds or cultures that was responsible for the development of modern science in Europe, these scholars argued, but easily identifiable economic and political interests of European cultures at that moment in time. Rereading such work through the lens of post-Kuhnian and postcolonial science and technology studies, one can see such externalist history arguing that systematic knowledge, not only systematic ignorance, can be generated by the local character of modern sciences and technologies.

Boris Hessen, a young Soviet historian murdered in the Stalin purges of the 1930s, was one of the first of these externalist historians. He argued in 1932 that the theoretical problematics of early modern science—specifically those that Newton addressed—were just those that needed to be solved in order for the industrial takeoff in Europe. It was emerging capitalism that stimulated the development of Newtonian mechanics.[15] Hessen's argument was not the psychological one that Newton intended to assist the industrial takeoff by solving such problems; the argument was not that he was consciously motivated to do so. Nor did he argue the materialist, determinist position that Newton's historical moment dictated in any sense what scientific problems Newton would pursue. Rather, Hessen's point was that certain issues were "in the air" at the time; they were especially interesting to inquisitive minds such as Newton's at that historical moment and location and not to thinkers earlier or elsewhere. What made Newton such an important figure in the history of science was precisely the *subsequent* importance of the issues he addressed. It was in this sense that Newton's highly theoretical work thus was fully part of his particular historical moment, Hessen argued.

In the 1950s, sociologist Edgar Zilsel, a member of the Vienna Circle, argued that the experimental method that was regarded as the crucial feature giving rise to modern science could not have been developed in slave societies or aristocracies where there existed a severe class division between head and hand laborers. Experimental method required training in both manual and intellectual labor. However, the aristocracies in such class-based societies assigned manual labor to people of low status—serfs, peasants or slaves, and they prohibited or otherwise put out of reach the possibility of the intellectual development of serfs, peasants or slaves. Manual labor was devalued by the educated classes precisely because it was associated with the low social status of those who did it. Thus, such societies effectively segregated from each other the two kinds of activity that experimental method so effectively brought together. It took the beginnings of the eventual breakdown of the medieval European class system, Zilsel argued, to permit the emergence of various kinds of artisans—painters, sculptors, architects, navigators, glass and instrument makers, cartographers, physicians—who combined the best intellectual training of their day with manual know-how.[16] Indeed, other historians have identified how little actual experimental work was done by the scientists whose names are associated with such experiments as late as even Clark Maxwell in the nineteenth century. Maxwell employed a series of servants to carry out the experiments associated with his name.[17] Yet other historians have focussed on the development of a group of landowners with interests in improving their land and on other possibly significant aspects of European societies that could account for the shift from feudal to modern social formations and their distinctive scientific and technological projects.[18]

Many issues raised by these accounts take on new meanings in the context of postcolonial accounts. Of course the externalist accounts remain controversial, for they deny that the seeds of scientific and technological growth are to be found as completely in the features of scientists' minds and their research procedures as internalists think the case. And their externalism has proven to be excessive; they have only begun to grasp how the social order "gets inside" the cognitive content of scientific claims and processes, as the subsequent northern science studies have been able to show. However, there is another problem with them even for those who find the internalist arguments unconvincing. These kinds of accounts still are shaped by unjustified eurocentric diffusionist assumptions. They ignore the effects of European expansion on the emergence of modern science and technology in Europe, and their observational field does not include scientific and technological traditions other than those upon which the traditional internalist histories focussed. From the perspective of later strains of postcolonial science studies, Marxian explanations that remain unmodified by postcolonial history and theory remain eurocentric insofar as they persist in identifying the sources of the growth of European science and technology as lying entirely within Europe.

Joseph Needham's histories of Chinese science and technology that began appearing in the 1950s hold an ambivalent though important position in this kind of account of the shift from diffusionist eurocentric to postcolonial science and technology histories. On the one hand, they ultimately must be located on the side of the internalism assumptions. They are internalist in that they still treat what Needham identifies as the central features of modern science as unique and internal to European science.[19] Moreover, focussing as Needham does on China, he does not fully appreciate the immense effects that the Voyages of Discovery to the Americas especially had on the development of modern sciences in Europe. They lack these two central features of the postcolonial histories.

On the other hand, Needham shows the remarkable accomplishments of Chinese sciences and technologies, and the many borrowings (diffusions!) from them into European sciences. Again and again he criticizes the distinctly eurocentric, racist, and otherwise unobjective accounts European and U.S. historians and philosophers make of Chinese sciences and technologies and their blindness to the significant

European and Christian dimensions of early modern sciences and technologies. Needham succeeds in producing a more balanced accounting of the relative accomplishments of Chinese and European sciences, highlighting some of the origins of the latter in the former. Indeed, Needham has been unjustly neglected by today's historians of science.[20] What is important for the argument here is that in analyses such as Needham's can be found some tributaries of the postcolonial histories that have subsequently developed.[21]

c. World systems theory. Something had been not completely visible even to the Marxian externalists' historical gaze, and that was the full significance of European expansion, and especially the Voyages of Discovery, for the history of both European and non-European scientific and technological traditions. However, this relation between European expansion and the Voyages of Discovery had not always been so hidden; it had only relatively recently disappeared from European consciousness. Blaut points out that "the thesis that industrial development in Europe depended in many ways on colonial processes was widely accepted in the eighteenth and early nineteenth centuries."[22]

The first resurfacings of this awareness appeared in English in the 1930s and 1940s when a number of colonial scholars began developing arguments that European expansion had enabled the industrial takeoff in Europe and the de-development, both intentional and accidental, of cultures into which Europe expanded. For example, Indian scholars argued that

> a highly developed Indian cotton textile industry not only provided some of the new technology for Britain's industry, particularly in dyeing, but also had to be forcibly suppressed by Britain—in a process which some Indian scholars call 'the de-industrialization of India'—in order to allow the British industry to develop in the late eighteenth and nineteenth centuries.[23]

Moreover, two West Indian scholars, C. L. R. James and Eric Williams, began to show how the slave trade and slave-based industries "were crucial causal forces in British and French industrialization."[24] They used the lens of the Marxian model of European class relations between bourgeoisie and proletariat to look at the global labor relations that European nations became able to control through their overseas expansionist projects.

Themes of these early critiques of diffusionism continued through the vigorous discussions in one stream of northern history of science and technology in the 1960s and 1970s. George Basalla's proposal for a three-stage model of diffusionism met with challenges from different parts of the world about the relations between European imperialism and the flourishing of modern sciences and technologies in Europe.[25]

During this period interest in causal forces moving in the other direction developed in world systems theory's concern, following the lead of the Indian and Caribbean colonial historians, to explain how the development of Europe also de-developed much of the rest of the world. Andre Gunder Frank and Immanuel Wallerstein in the late 1960s and early 1970s began to explore the concepts of "metropolis-satellites" and "core-periphery," respectively, as ways of developing antidiffusionist accounts of global history. As Frank argued,

> Historical research demonstrates that contemporary underdevelopment is in large part the historical product of past and continuing economic and other relations between the satellite underdeveloped world and the now developed metropolitan countries. Furthermore, these relations are an essential part of the structure and development of the capitalist system on a world scale as a whole.[26]

Blaut summarizes this fifty-year-old argument about the modern era as it contributed to the development of the emerging postcolonial histories in the following way:

> Capitalism arose as a world-scale process: as a world system. Capitalism became centrated in Europe because colonialism gave Europeans the power both to develop their own society and to prevent development from occurring elsewhere. It is this dynamic of development and underdevelopment which mainly explains the modern world.[27]

This is the kind of historical map on which postcolonial science studies are now locating their questions about why modern sciences emerged in Europe. On such a map can be located historical and contemporary ethnographic studies of the science and technology traditions of non-western cultures, and their contributions to the development of modern sciences.

   d. Anti-eurocentric comparative ethnoscience studies and the non-European origins of modern sciences and technologies. Of course there is a long history of eurocentric studies of other cultures' traditional ways of thinking about the natural world. But these putative knowledge systems— "ethnosciences"—never produced real knowledge of the principles of nature's order. In the eurocentric accounts, they were presented as anthropomorphic or religious beliefs about nature, full of superstitions, magical beliefs, and dangerous patterns of ignorance. On the margins of such eurocentric studies were claims that "high cultures," such as those of Islam and far-eastern Asia really did make a few contributions to Europe's history.

   In the postcolonial science studies of today can be seen two themes that began to appear out of and yet also in opposition to the eurocentric standpoint of these literatures. A few ethnographers of non-European science and technology traditions began calling for treatments of all cognitive systems as on an epistemological par, thereby refusing to recognize the usual epistemological distinctions between real knowledge and mere local belief.[28] These studies did not deny that some belief systems were able to achieve more powerful effects than others, of course. Instead, they denied only that the causes of such successes were to be found in the purportedly purely internal epistemological features of modern scientific processes—their inherent rationality, unique logic of justification, universal language, objectivity-achieving method, and so on. Thus, the challenge here is not to the existence of distinctive forms of rationality, logics of justification, distinctive languages and methods, and so on, in modern science, or to their usefulness to the advance of knowledge (when that is what they do). What they challenge is, rather, the status of such epistemological features as purely internal on the grounds that they have escaped the kinds of distinctively cultural character that marks all other human products. These rationalities, logics, languages, and methods, too, are fully historically and culturally local, the comparative ethnoscience scholars have argued. (We return to their arguments in later chapters.)

This "epistemological equality" perspective in the ethnographies and histories permitted a fuller appreciation of the rich scientific and technological traditions of non-European cultures—the second theme. This is so even though such accounts often included residual, comparatively muted, eurocentric caveats in the course of attempting to account for the rise of modern sciences in Europe and decline of other scientific and technological traditions. These two themes also opened up a space for raising questions about how other cultures had contributed to the development of modern sciences and technologies in Europe, and how such borrowing had been obscured in the conventional accounts.[29] European sciences have been greatly enriched by borrowings from other cultures, subsequent histories showed. Of course, most people are aware of at least a couple of such examples. However, the borrowings have been far more extensive and important for the development of modern sciences than the conventional histories reveal. It is not only the so-called "complex" cultures of China, India, and others in east Asian and Islamic societies that have provided resources for European sciences and technologies, but also the so-called "simpler" cultures of Africa, pre-Columbian Americas, and others that were encountered by Europeans during their Voyages of Discovery.

For example, Egyptian mystical philosophies and premodern European alchemical traditions were far more useful to the development of sciences in Europe than is suggested by the conventional view that these are only irrational and marginally valuable elements of premodern thought.[30] The Greek legacy of scientific and mathematical thought was not only fortuitously preserved but also developed in Islamic culture, to be claimed by the sciences of the European Renaissance.[31] Some knowledge traditions that were appropriated and fully integrated into modern sciences are not acknowledged at all. Thus, the principles of pre-Columbian agriculture, that provided potatoes for almost every European ecological niche and thereby had a powerful effect on the nutrition and subsequent history of Europe, was subsumed into European science.[32] Mathematical achievements from India and Arabic cultures provide other such examples.[33] The magnetic needle, rudder, gunpowder, and many other technologies useful to Europeans and the advancement of their sciences (were these not part of scientific instrumentation?) were borrowed from China.[34] Knowledge of local geographies, geologies, animals, plants, classification schemes, medicines, pharmacologies, agriculture, navigational techniques, and local cultures that formed significant parts of European sciences' picture of nature were provided in part by the knowledge traditions of non-Europeans. ("We took on board a native of the region and dropped him off six weeks further up the coast," the voyagers generally report.) Summing up the consequences for modern sciences of British imperialism in India, one historian points out that in effect "India was added as a laboratory to the edifice of modern science."[35] One could say the same for all the lands to which the Voyages of Discovery and later colonization projects took the Europeans.[36]

The eurocentric accounts habitually claimed the Greek legacy as European, "unpolluted" by non-European elements during its sojourn in Alexandria. Thus, they denied the way the spread of Islam placed Aristotle and his world in the legacies of many Asian and African cultures. There was the questionable tendency to project today's European borders and presumed cultural identity back into the ancient Greek world, though Aristotle's Greece was part of a Mediterranean world that included the Near East and Northern Africa, and excluded all peoples lying to the north of Greece. The invention of the European miracle, the Dark Ages, and the scientific revolution all worked to obscure and deny the non-European origins of early modern sciences and technologies, as we saw earlier.

Such accounts show that modern science already is multicultural at least in this sense that elements of the knowledge traditions of many different non-European cultures have been incorporated into it. Of course, there is nothing unusual about such scientific borrowing. It is evident in the ordinary, everyday borrowing that occurs when scientists revive models, metaphors, procedures, technologies, or other ideas from older European scientific traditions, when they borrow such elements from the culture outside their laboratories and field stations, or from other contemporary sciences.[37] After all, a major point of professional conferences and international exchange programs, not to mention "keeping up with the literature," is to permit everyone to borrow everyone else's achievements. Without such possibilities, sciences wither and lose their creativity. What is at issue here is only the eurocentric failure to acknowledge the origins and importance to modern sciences and technologies of these borrowings from non-European cultures, and also thereby to trivialize the achievements of their scientific and technological traditions. Postcolonial histories have enabled a more objective understanding in this respect of the causes of growth in European sciences and technologies. They also draw to our attention the great achievements of other traditions and the tragedy of their decline and extinction in the many places that this has occurred.

e. Is Third World "development" merely the continuation of colonialism by other means? Yet one more major tributary into contemporary postcolonial science and technology studies has been the criticisms of the so-called development policies that were instituted by Europe and the United States after World War II in the First versus Second Worlds' attempts to secure for themselves the allegiance of the "unaligned nations"—a group that would become the Third World in the Cold War rhetoric shaping this development policy. Designed by national and the new international agencies of the United Nations, development policy was largely organized and implemented out of the same buildings and by the same personnel that had the month before been organizing and implementing colonial administration, as the postcolonial critics point out.

Shortly after the implementation of development policies, it became clear that they were actually worsening the conditions of the majority of their recipients, namely the world's poorest peoples, who were already the most economically and politically vulnerable and thus not in a position to get their assessments of their needs heard in the development policy centers. Development was continuing to move natural and social resources from the "have nots" to the "haves," in the parlance of the period, and thus largely creating de-development and maldevelopment in the Third World while further "developing" northern economies. Slowly an analysis began to emerge demonstrating that a major problem was the conceptualization of development in purely economistic terms—as economic growth. Northern science and technology policy were thus implicated in the development failures since economic growth had meant the transfer to the south of the industrial models thought to have been responsible for the growth of the European and U.S. economies in the eighteenth, nineteenth, and early twentieth centuries. Scientific and technological knowledge was thought central to the development of these industrialization models. Indeed, development was often referred to as science and technology "transfer" from the North to the South. However, the North's conception of economic growth in the South in fact succeeded largely in creating small middle classes there that were allied to northern policy makers, widespread environmental destruction, and extensive interference in and destruction of southern cultures by the northern ones. So-called development was largely the continuation of colonialism by other means, these accounts claimed.[38]

These development critiques continue in the context of the late-twentieth-century historical, philosophic, and political themes developed in other contexts in the other tributaries to contemporary postcolonial science and technology studies. Later chapters will pursue in greater detail these arguments about the failures of development policy.[39] Like the other tributaries, these accounts open up space for new questions and new dialogues about the scientific and technological traditions of modern Europe and of other cultures around the globe.

3. Conclusion. This chapter's brief sketch of some of the important origins of postcolonial science and technology studies today no doubt leaves out some literatures and themes, and thus has the effect of overemphasizing others. This arena of science and technology studies today is a diverse one of intertwined research, scholarship, and policy work, with conflicting disciplinary, theoretical, and political assumptions—even among researchers in neighboring offices—insuring lively ongoing disputes. This field is very much in the making; how it will get composed cannot be either shaped or predicted from any single disciplinary, theoretical, or policy location within it. It is too important a field of knowledge production to too many groups around the world, and for too many conflicting reasons, to be controllable by any subset of them. This chapter has not been intended as a comprehensive review of some half-century of analyses that are actively and continually being reevaluated; it has not been intended as a detailed photograph of this complex history.

Rather, this account has set out to identify some of the undoubtedly major sources of what is by now an increasingly familiar conceptual map on which valuable understandings, both new and older, of human scientific and technological achievements can be located in what were recently unforseen and, indeed, even unimaginable new relations to each other. Postcolonial science and technology studies present us with a different picture of the past, present, and future possibilities than were visible in the older, eurocentric "tunnel of time" accounts of European and other scientific and technological traditions. It brings into focus some objects, events, and processes that were obscured or, in some cases, denied in older views. The usefulness of such an organization also moves into the background or entirely off the stage other phenomena and concerns that were of central interest in the eurocentric accounts.

Just as the questions "where did the crystal spheres go?" and "how will God find the good and the sinners if we are no longer at the center of his universe?" had no satisfying answers after Copernicus and Galileo, so, too, some of the older empirical and philosophic questions central to the eurocentric, internalist accounts simply have no answers within these postcolonial science and technology studies.

These studies also open up a gap between the dominant epistemologies and philosophies of science of the modern West and the ones needed to account for the history of knowledge production in Europe and elsewhere revealed by the postcolonial histories. What theories of knowledge will account for this kind of history of the rise, flourishing, and often decline of cultures' patterns of systematic knowledge and systematic ignorance?

The next chapter looks more closely at the relationship between European expansion and the emergence of modern sciences and technologies in Europe. This is a series of encounters between cultures where one can see all the themes of this chapter joining to open up conceptual space for new questions about the epistemological relationship between science, modernity, and European expansion.

3

Voyages of Discovery:
Imperial and Scientific

> The rise of modern science and the colonial expansion
> of Europe after 1492 constitute two fundamental and
> characteristic features of modern world history. . . .
> The story is a dual one. One of its aspects concerns how
> science and the scientific enterprise formed part of
> and facilitated colonial development. The other deals
> with how the colonial experience affected science and
> the contemporary scientific enterprise.[1]

1. A standpoint on the voyages. Historian James McClellan names an issue that has directed one important tributary of the postcolonial science and technology studies that have been developed since the 1960s when formal European colonial rule began to decline. How were European expansion and the development of modern sciences in Europe—these two great processes of modernity—related to each other? Has the plausibility of the prevailing epistemologies and philosophies of science depended upon the assumption that there are no significant causal relations between the successes of European expansion and of modern sciences? If these two processes did in fact play a major role in facilitating each other, how should our understandings of the rationality and objectivity of modern sciences, and of the relationship between knowledge and politics, change? How should we assess the rationality and objectivity of the conventional philosophies and social studies of modern sciences and technologies that could not see this relationship?

Such questions could not even be asked from the perspective of the older "internalist" approaches that attributed modern science's successes solely to culture-free epistemological features that are internal to the nature and processes of the sciences. Moreover, the "externalist" histories of science have largely shared with the internalist ones the view that there is a pure, value-neutral cognitive core to modern science that is unaffected by the historical and sociological accidents of how scientific and technological knowledge in fact are produced. However, the convergence of several strains of postcolonial science and technology studies that have arisen since World War II has cleared a historical and conceptual space that makes it possible to pose such questions in ways that can now look interesting and important. (The internalist, externalist, and postcolonial accounts were described in the preceding chapter.) Postcolonial studies clear space for asking kinds of questions that do not fit within the conceptual frameworks and fundamental assumptions of either the older internalist or externalist accounts. However, as pointed out in earlier chapters, these studies converge in important respects with the post-Kuhnian philosophies and science studies produced since the 1960s that try to show the integrity of modern sciences with their historical moment. This, too, is the project of postcolonial science and technology studies. The postcolonial studies use the resources of the postcolonial histories and geographies to show the changing relations of both European and other cultures' knowledge traditions with the global economic, political, and social changes brought about by European expansion from 1492 through so-called Third World development today.

From the standpoint of postcolonial studies one can gain more accurate and comprehensive descriptions and explanations of the science and technology traditions of non-European cultures. Equally important, however, is that we can also gain more objective understandings of just what has been responsible for the successes of the European traditions and for their limitations. Philosophic, historical, and sociological theories of what does and should count as the most reliable kinds of knowledge, and of how to get more of it, should be able to account for what are regarded as the best examples of such knowledge in the past. What would be the point of a theory of knowledge that could not account for what its adherents took to be the greatest knowledge achievements that humans have managed to produce? Thus, the new histories of science and technology, which were themselves made possible by the beginnings, from the 1950s on, of the end of European colonialism, offer food for philosophic, historical and sociological thought about how scientific knowledge has been and can be generated.

The year 1492 is a convenient marker for the beginning of European "Voyages of Discovery" that headed out not only toward what turned out to be the Americas, but also around the Cape of Good Hope toward the east coast of Africa, India, China, the rest of southeast Asia, and, several centuries later, Australia and New Zealand. This chapter focuses on the role in particular that the Discovery of America, as it used to be called, played in facilitating the development of modern sciences and technologies in Europe. While what happened in other parts of the world is an important part of this story about Europe and the Americas—as one would expect within the interactionist global history framework developed by postcolonial studies—the account here will "start from the Americas." Interesting additional perspectives on this Americas story and on the development of modern sciences and technologies in Europe would emerge if one started out telling the postcolonial account of the Discovery of America from the standpoint of the peoples in Africa, China, or India whose lives were radically changed by processes of European expansion in the Americas. However we shall focus solely on this direction of European expansion in order to sketch out more fully some of the themes only outlined in the preceding chapter's analysis.

The following sections will look at the two-way causal relations between the development of modern sciences and technologies in Europe and three aspects of European expansion: Europeans' attempts to get to the Americas and establish "little Europes"; the development of colonial sciences "in" these new worlds; and the de-development of the indigenous scientific and technological knowledge traditions of the peoples of the Americas. (The philosophic questions these accounts raise will be pursued in later chapters.)

2. Getting there and establishing "little Europes." What scientific, technological, and social information was needed in order for Europeans to get to and then survive in other parts of the world? As environmental historian Alfred Crosby puts the question in his prize-winning study, what was needed for Europeans to establish "little Europes" in other parts of the world?[2]

Crosby shows how Europeans needed to figure out the system of winds and tides in the North and South Atlantics that would permit them not only to sail to the Americas but—a much more difficult problem—to get back to Europe. This knowledge would also enable them to make the difficult trip south around the Cape of Good Hope and into the Indian Ocean. They needed better maps of the shores of the Atlantic and of the lands in which they would try to set up hospitable living conditions. They needed better navigational methods, including charts of the heavens as seen from the southern hemisphere. They needed ships capable of such lengthy and difficult voyages. They needed provisioning stations en route to their destinations to enlarge their food supplies. The Canary Islands provided their first such station for all of these trips. A botanical garden in Capetown, South Africa is on the site of the provisioning garden established by the East India Company some five centuries ago. It is one of the many such gardens surviving today that originated as just such a source of food for the voyagers.[3]

Once they got to the Americas, the Europeans needed to learn how to survive there amid the unfamiliar and often hostile vegetation, animals, diseases, peoples, terrains, and, to a lesser extent, climates. As

Crosby points out, the temperate zone climates of the Americas were already familiar to Europeans, and the key to their ability eventually to establish the meat and produce production that could find a ready market in the expanding populations in Europe. Seeking riches, they needed to find labor to mine the gold and silver and later to work on the agricultural plantations, forestry, or manufacturing industries that they established in the colonies. The Europeans had to be able to establish European econiches where they could either grow familiar foods and use their familiar ways of gaining shelter, travel, pharmacologies, and the other necessities of life, or learn how to adapt local resources to survival purposes.

Initial attempts to settle in the Americas were unsuccessful. Various attempted communities were abandoned or lost on the northernmost coasts of the Americas—so much more inhospitable even today than their counterparts at the same latitude in Europe. Even colonization attempts much further south, as in Jamestown, Virginia, for example could not at first manage to survive. The Americas were relatively welcoming sites, however, for such "little Europes" compared to some of the other places in which the Europeans tried to settle.

The ease of establishing such colonies was made possible by two main circumstances in their favor: a small superiority in European firearms and, even more important, the susceptibility of the American indigenes to European diseases. Geographer J. M. Blaut argues that these were the only two significant advantages on the side of the Europeans, contrary to standard eurocentric claims attributing the success of European expansion to the superiority of European character and culture. Prior to 1492, the conditions of late medieval societies around the globe were more or less similar, he points out. Their modes of production, class structures, systems of spatial exchange, patterns of urbanization, and stages of scientific and technological development were not significantly different from each other. In all of these societies the feudal systems began to decay and collapse. All of them showed signs of change toward commercialized agriculture and rural capitalism.[4] But what permitted Europe to forge ahead was its sudden access to the wealth of the Americas, access that was eased by a small superiority in firearms and the indigenes' susceptibility to the Europeans' contagious diseases.

> The technological gap was not so great that it could by itself bring military victory—after the initial battles—against American armies that were vastly larger and would sooner or later have adopted the enemy's technology. America is a vast territory, and in 1492 it had a very large population, numbering at least 50 million people and conceivably as many as 200 million, a goodly proportion of these people living in state-organized societies with significant military capability. . . . Moreover the superiority of the Spaniards' primitive guns was not really very great when compared

with the Americans' bows and arrows. . . . History went in a different direction because of the incredibly severe and incredibly rapid impact of introduced diseases. (184)

Many of these diseases had become epidemiologically significant in human history only during the Agricultural Revolution that had long ago occurred in Europe and Asia. Though there were certainly agricultural societies in the Americas, they were more sparsely distributed, and they were located in different environments than those in which the European diseases had developed. The first few Europeans who landed in the West Indies infected the indigenes they encountered, and these diseases spread like the proverbial wildfire through the local populations. By the time the conquistadors arrived a decade after Columbus, the populations of the cultures they met had already been reduced by the European diseases from 50 percent to, in one case, 100 percent. During the sixteenth century, "probably 90 percent of the population of central Mexico was wiped out by disease, and three-quarters of the entire population of America."[5] The familiar eurocentric myth of the emptiness of the lands the Europeans conquered looked plausible to the Europeans by the time large numbers of them arrived in the Americas.[6]

Thus, the Europeans needed and developed many kinds of scientific information and technological know-how in order to get to other parts of the world and to survive there. Some of this they developed on their own, spurred on by the expected benefits of the voyages and enabled by the new aspects of nature's order that they encountered in their travels. Other kinds they borrowed from the indigenes they encountered in these initial attempts to secure a landing place in other lands. And, obviously, different fields of science benefitted in different ways and to different degrees from the voyages—as they would from colonization.

To arrive and survive were great achievements. However, to thrive required additional knowledge and technologies. The "conquest of the Americas" was especially central to the subsequent flourishing of European populations and their culture both in Europe, and other parts of the world for it was the Americas that provided two key resources. They provided the silver and gold with which subsequent voyages, the enlarged trading projects, and the establishment of colonies could be financed. And they provided the seeds that would transform the nutrition of Europe and many other parts of the world, and that could be developed as commercial crops in colonial plantations established in the Americas, India, Africa, and other parts of the world. And colonization would, in turn, mightily advance the growth of Europeans' knowledge about nature's order.

3. Colonial science. A small number of examinations of European sciences in colonial contexts have appeared in the last decade or so. These have had diverse focuses. For example, one examines the value to both European science and to European expansion of the botanical gardens that Europeans established in Europe and in the places they visited or settled, such as the Kew Gardens outside London and the East India Company's botanical garden in Capetown.[7] Another looks at one of the few scientific societies that existed in the European colonies, the "Cercle des Philosophes" in eighteenth-century Saint Domingue.[8] A number examine the historical shifts in how European-directed scientific work was focussed and how it was organized within one colony over time, such as the British science and technology projects during the two-hundred-year occupation of India.[9] Such a focus includes the development of distinctively colonial scientific fields such as "tropical medicine." And others analyze different aspects of the relationship between the development of modern sciences in Europe and European expansion.

Five arguments in these literatures are of special interest to our project here. The colonists' science projects were, first and last, for maintaining Europeans and their colonial enterprises in those and other parts of the world. They were designed especially for increasing the profit Europe could extract from other lands and maintaining the forms of social control necessary to do so. Second, mainland European science cannot be isolated or immunized from the effects of the colonial and other expansionist projects. The establishment of European colonies resulted in immense contributions to the growth of science in Europe, these accounts reveal. Third, colonial activities in one part of the world enabled the establishment of political, economic, scientific, and technological resource flows toward Europe not just from that particular part of the world but also from others. Colonial projects in any one area of the globe permitted Europeans exponentially to increase their ability to organize a disproportionately large share of the world's supply of human and natural "energy." Fourth, colonialism frequently intentionally, but often also unintentionally, systematically destroyed local scientific and technological traditions and resources the better to benefit Europeans. Finally, post–World War II so-called development policy in effect continues such processes: development is largely "colonialism by other means." Let us explore further the first four of these claims. (The fifth was outlined in the last chapter and will be the focus of analyses in several later chapters.)

Sciences "at the vanguard of colonialism." The European colonial science projects "marched at the vanguard, not at the rear guard, of colonialism," at least for some European nations, as McClellan puts the point in reflecting on the role of eighteenth-century French scientific projects in the Caribbean.[10] It is worth quoting from several of these accounts to get a clearer sense of this argument. Here is McClellan:

Science and organized knowledge did not come to Saint Domingue as something separate from the rest of the colonizing process but, rather, formed an inherent part of French colonialism from the beginning. In other words, the French did not colonize Saint Domingue and then import science and medicine as cultural afterthoughts. French science and learning came part and parcel with French colonialism, virtually as a 'productive force.' Because they were already so institutionalized in the culture and state apparatus of France in the eighteenth century, science and medicine played—or seemed to play—important roles in the development of French West Indian colonial interests. (7–8)

Research topics were selected not because they were intellectually interesting (though no doubt they often were), but but to solve colonialism's everyday problems:

[C]olonial researchers took up problems not on account of any internal scientific tradition but because of the exigencies of the local context in Saint Domingue. The Cercle des Philadelphes and other colonial medical investigators undertook extensive studies of tetanus, for example, first and foremost because the disease posed a serious threat in Saint Domingue. The research done in the colony on poisons and their effects is likewise to be ascribed to colonial anxieties over poisonings by slaves. Similarly, progressive colonists studied coffee and the indigo plant not simply as students of botany but because coffee and indigo constituted important commodities in the economy of the colony. (295)

One cannot escape the conclusion that important parts of European organized scientific research were fundamentally in the service of establishing and maintaining colonialism and slavery:

Science could investigate causes of tropical diseases, determine Saint Domingue's precise location, identify useful flora and fauna, and explore potential new avenues for colonial development, particularly promising economic enterprises. . . . Without doubt, French science and medicine bolstered slavery and the slave system in Saint Domingue. . . . Directly and indirectly, then, scientific expertise deployed to build the colony constituted support for slavery and the slave system, and the theme of science and slavery in old Saint Domingue provides a sober reminder that in some instances at least science has served 'unfreedom' and human oppression. (8–9)

R. K. Kochhar, a historian from the Indian Institute of Astrophysics in Bangalore, writes similarly about British science in India:

the introduction and growth of modern science in India was with a view to serving the colonial interests. Thus the British-sponsored science, by the very reason of its existence, was field science. Geography, geology and geodesy, botany and zoology, archeology, medicine and even astronomy— all these stemmed from the physical and cultural novelty of India. This science was colonial in the sense that its agenda was decided on grounds of political and commercial gain.[11]

And historian Lucille Brockway argues that

> the cultivation and development of a quinine industry was an arm of British empire in Africa. Because of its tropical diseases, Africa had been known as "the white man's grave," and trade had been conducted from coastal trading posts, contact with the hinterland being left to African middlemen. West Africa was the deadliest malarial environment in the world, with *falciparum* malaria, the worst form of the disease, hyper-endemic in many localities.[12]

Brockway reports that from 1817 to 1836, the death rates among British military personnel recruited in the United Kingdom were 483 per thousand from disease alone in Sierra Leone, and 668 per thousand from all causes in the Gold Coast between 1823 and 1826. This was compared to twelve per thousand in the Eastern Frontier District of South Africa, seventy-five in Ceylon and 130 in Jamaica.[13]

The postcolonial science and technology histories record case after case where scientific research clearly was not intended to increase "human" freedom and general social welfare for the peoples Europeans encountered. Instead, its achievement was to increase the share of social benefits that accrued to small elites, those colonists who survived and prospered, and even more so, to their sponsors in Europe who were enriched by the gold, silver, platinum, and plantation profits that colonization could produce. For the great majority of participants in European expansion—willing or not—the Europeans' scientific and technological projects delivered increases in mortality rates and in human bondage.

In these postcolonial accounts, such scientific projects are not a bizarre aberration that can be distanced by attributing it only to the technologies that even evil or misguided people can develop from pure science, or to other political misuses and abuses of sciences. It is not that scientists cannot control the destinies of sciences in the everyday world. Instead, one can see how these local projects of colonialism and imperialism played a central role in the development of modern sciences and technologies in Europe at the time they occurred, rather than in some other society elsewhere around the globe. If sciences marched at the vanguard of colonialism, so, too, did colonialism march at the vanguard of at least some European sciences.

Colonialism for sciences. The voyages and establishment of European colonies made an immense contribution to the growth of science in Europe. Kochhar points out that "the navigational needs of the traders acted as a great incentive for development of science in Europe." For example, "the best scientists of the time applied their minds to 'discover the longitude,'" and to that end established observatories in Paris (1667) and Greenwich (1675). Professors at British universities, such as at Gresham College in London, "took up navigational problems in the national interest, e.g., Henry Briggs (1560–1631), whose introduction of logarithm to the base 10 greatly simplified mathematical calculations."[14] In the 1730s sea quadrants were invented in England.[15]

The different ways that scientific and technological work was organized in the European nations shaped the diverse consequences that colonial scientific projects would have on the European mainland sciences. As McClellan writes about the Saint Domingue colony, French scientific institutions had a strong hand in colonial development since French science and medicine were heavily state organized and institutionalized.

> ... [T]he empire of French science expanded along with the rest of the French colonial empire in the eighteenth century, and in the end, institutional ties linking Saint Domingue and metropolitan France indicate that colonial science in Saint Domingue, while on the periphery, was anything but peripheral.[16]

In a fortuitous phrase (cited in earlier chapters), Kochhar characterizes the consequences for modern sciences of British imperialism in India by noting that in effect "India was added as a laboratory to the edifice of modern science."[17]

Sciences for a European global economic, political, and scientific network. The establishment of colonial plantations and manufacturing permitted Europeans to organize exponentially the world's supply of human and natural "energy," as Brockway puts the point. From the beginning, it was this project of organizing the world's energy that enabled the Europeans to take the political lead. Trade with the East was linked to resource extraction from the Americas.

> At first the Europeans had nothing but their firearms to trade for the luxury goods of the East, for Europe produced little else that Asia wanted. ... The lack of saleable commodities to trade with the East caused a drain of gold and especially silver throughout the sixteenth and seventeenth centuries which was only partially offset by the earnings of Portuguese, and later Dutch, ships in the inter-Asian carrying trade, from the Persian Gulf as far as China and Japan. ... a Florentine traveler ... estimated that the Chinese extracted from Portugal and Spain "more than a million and a half ecus a year in silver, selling their goods and never buying anything, so that once the silver gets into their hands it never leaves them."[18]

While some of this silver went directly from Mexico to China, most of the New World's precious metals went first to Europe, and then was used to settle the Europeans' trade deficits to the Far East. "Intercontinental trade in the sixteenth and seventeenth centuries consisted essentially of a large flow of silver which moved eastward from the Americas to

Europe and from Europe to the Far East, and a flow of commodities which moved in the opposite direction."[19]

Seed and plant transfers between the Old and New Worlds after 1492 greatly contributed to the development of the West and its "political and economic expansion into the rest of the world in search of food supplies and raw materials."[20] As new food staples increased every continent's populations, industrializing Europe could make good use of this increased labor pool. Moreover, the increased populations in "old agrarian empires of Asia" very shortly found themselves laboring to serve European projects. It was Europeans who predominately organized and directed the various plant exchanges between the two hemispheres, using plantation systems of coerced or servile labor. It was Europeans who "made all the rules and reaped the profits, thereby accumulating capital for the development of their industrial societies while deforming the societies which supplied the raw materials and labor."[21] In subsequent eras, the organization of plant transfers became more formalized, and by the middle of the nineteenth century, it was the botanic gardens that began to initiate and sponsor the collecting voyages. Moreover, they had by then learned how to improve and adapt plants for commercial production. Competition between European governments to "establish botanical monopolies and to break the monopolies of their rivals . . . was a spur to plant development."[22]

Introducing her case studies of the centrality of Kew Gardens in England to the development of the quinine (cinchona), rubber, and sisal industries, Brockway points out how nineteenth-century Europe was

> achieving a global dominance, extracting and mobilizing the energy of the world for its own purposes. In each of my three case studies, a protected plant indigenous to Latin American was transferred by Europeans to Asia or Africa for development as a plantation crop in their colonial possessions. Brazil, Mexico, Columbia, Peru, Ecuador and Bolivia each lost a native industry as a result of these transfers, but Asia acquired them only in a geographical sense, the real benefits going to Europe.[23]

Brockway's account provides just one example of the way that colonialism and modern science worked hand in hand to organize European power through globally linking raw materials from one area of the globe with the labor and skills of cultures in other parts of the globe to produce benefits for Europeans.

European expansion's destruction of competitive scientific and technological traditions. European expansion delivered another benefit to the ascending global status of European scientific and technological achievements. The knowledge traditions of the cultures Europeans encountered did not just happen to lag behind Europe's great scientific and technological advances. Instead, these traditions were de-developed through processes of European expansion—sometimes intentionally, sometimes accidentally.

The postcolonial accounts identify at least six main ways in which this de-development of non-European scientific and technological traditions occurred in the Americas as well as in other cultures into which European culture expanded. First, European expansion extracted raw materials from other cultures, ranging from gold and silver through botanical specimens. Such materials then supported the growth of European societies and their modern sciences and could no longer provide resources for their cultures of origin. Second, it extracted labor from these cultures, from the Central American and African gold and silver mines, to the agricultural production on plantations in the New World, Africa, and Asia, and the manufacturing of textiles and other goods, especially in Asia. Europeans moved raw materials to the labor, or, in the case of the plantations and mines of the Americas, labor to raw materials. In either case, indigenous labor was no longer available to support projects, including scientific and technological ones, that could be organized to benefit the non-European societies. Third, European expansion extracted local scientific and technological knowledge to benefit European culture. Local scientific and technological ideas for navigation, cartography, agricultural development, manufacturing, pharmacology, and other ways of dealing with local environments and peoples were borrowed from non-European cultures and incorporated into European projects, many of which were intended to decrease the economic and political power of the peoples Europeans encountered. Thus, the fruits of their own scientific and technological traditions were turned against them.[24]

Fourth, local industries and trades were destroyed. In some cases, this was intentional in order to make room for the European replacements, such as the destruction of Indian and, later, African dyeing and weaving industries to make way for European imports, and the destruction of indigenous subsistence multicrop agriculture to make room for European-controlled export monocrop agriculture. In other cases, it was simply a consequence of the general immiseration of these societies. Trade relations which had moved relatively freely between many of the Old World societies and between the cultures of the Americas, increasingly were rerouted through Europe during the expansion; they became the patterns of center/periphery relations. Fifth, European expansion decimated local populations. This occurred initially through the inadvertent introduction of diseases to which the indigenes had no immunity. It continued through warfare, the exportation of slaves, in Africa but elsewhere also, and the open hunting of indigenes by the Europeans (which remained legal in the western United States, Australia, and New Zealand until the end of the nineteenth century).[25] Contributing to their demise also was the general immiseration and destruction of life-sustaining conditions that the indigenous populations experienced under the conditions of labor for the Europeans, for example, in mining, agriculture, and manufacturing. Last but not least, European expansion devalued and destroyed local cultural traditions such that it was only with great difficulty, and in some cases, only in their own peripheries and diasporas, that they could sustain non-European cultural patterns or even elements of them within the newly introduced European ones.

Thus, European expansion contributed to the increasing gap between the achievements of European sciences and technologies and those of the cultures they encountered in two complementary ways. There were the active scientific projects Europeans organized to aid European expansion, and also there was the destruction of other knowledge systems either directly, or through the destruction of their human, cultural, and environmental resources.

Sciences for (de)development today. As indicated earlier, the postcolonial accounts also point out that many of these processes of de-development of other cultures and their scientific and technological traditions continue today under the label of "development." As described in the last chapter, development policy has replaced colonial policy, often in the same buildings and with the same managers that previously administered colonial policy. Now large international organizations, such as the International Monetary Fund, the World Bank, and other U.N. agencies administer policies that most of the time do not seem to be able to improve the lot of those vast majority of Third World populations that are the economically and politically most vulnerable—with women and children over-represented in this group. Recently, independent local Third World governments are usually willing and enthusiastic participants in bringing "modernity," "rationality," and what they perceive as the benefits of European science and technology into their societies.

These postcolonial science studies argue (as do I) that of course non-Europeans should have access to the benefits of international science and technology no less than do Europeans—that is not the question for postcolonial science studies. Rather, the question in these accounts is to what extent does so-called development reverse at all the direction of the flow of resources that colonialism established in a one-way stream from non-European countries to Europe. A second question is to what extent are the benefits of modern sciences and technologies that reach developing countries distributed below the level of their small middle classes and the already wealthy aristocracies. Scientific and technological change are inherently political, since they redistribute costs and benefits of access to nature's resources in new ways. They tend to widen any pre-existing gaps between the haves and the have-nots unless issues of just distribution are directly addressed. Under current development policies, the inequitable distribution of such costs and benefits are visible for all to see—at least among those who wish to see them. Our supermarkets and clothing labels show the continued flow of raw materials from South to North. We can see around us in our daily lives, on television, and in popular science magazines such ongoing processes as flows of workers, agricultural, manufacturing, domestic and professional—including mathematicians, engineers, and scientists—from South to North, flows of southern scientific and technological knowledge into northern projects; the flow of northern toxic wastes to the South; the clearing of land in the South for European agribusinesses; the flow of capital from South to North as the South futilely tries to keep up its interest payments to northern financial institutions on its so-called development loans . . . and more. Northern-designed development policies are not the only causes of these phenomena by any means, for local social inequalities in the so-called developing countries contribute to them also. Nevertheless, an important question for critical postcolonial studies is how development policy-makers in Europe and non-European societies have in fact addressed such distribution issues, and how they all can do so more justly in the future.

A major obstacle to such an assessment, as these accounts see it, is the faulty conception of science circulated for the last five centuries by scientists, policy makers, the media in the North, and their allies in the South.

> . . . [T]he role of science in converting knowledge into power for the core nations of the industrial world system is not . . . widely recognized, and its implications are resisted. In the postcolonial era this system continues to drain money, talent, and energy from the undeveloped countries in the form of information monopolies, patents, licenses, fees and other rents on technology, gross inequalities between buyer and seller in the tropical commodities markets, and underpaid labor. . . .
>
> Most developing countries are not in a position to generate their own scientific technology or to take a bargaining position in the face of the information monopolies held by the technically advanced countries. If the Third World is to achieve real development instead of modernization, if it is to alter its relationship to the power centers instead of becoming a more profitable area for them to exploit, or so desperately frustrated as to furnish an occasion for another world war, further questions must be asked about the role of scientific technology in the world power system.[26]

Brockway is concerned here with the important issue of how science converts knowledge into power. Of equal concern to the postcolonial accounts on which this study draws is how power converts interests and desires into science—in different ways and to different degrees for different scientific fields, of course.

Let us turn in concluding this chapter to identify some central theoretical issues for the history and philosophy of science that these postcolonial science and technology studies raise.

4. *New questions.* The tributary of "science and empires" studies examined here brings into focus more accurate and useful ways of thinking about the integrity of scientific endeavors with their historic moment—as historian Thomas Kuhn named the desired goal of the new kind of science studies he called for in the early 1960s.[27] Indeed, these postcolonial studies expand the geographic and historical map of Kuhn's project and that of the northern science and technology studies of the last three decades in three ways. Now one can begin to see the integrity of the cognitive content of successive eras of modern European sciences with global history instead of only with European history. Moreover, one can see this integrity from the standpoint not just of the European natives' traditional ways of thinking about it but also from the standpoint of other peoples that encountered the voyaging Europeans and their scientific and technological traditions. And one can begin to see also the integrity of the cognitive content of those other cultures' scientific and technological traditions with their own historical eras. These more expansive postcolonial histories reveal that the narrowly focussed geographic map of conventional histories has obscured important features of modern and other scientific and technological traditions.

However, they also show that the conventional periodization scheme has contributed to our systematic ignorance about both European and non-European sciences. The last chapter showed that the "European miracle" was no miracle at all; rather, its origins were perfectly visible in the processes of medieval societies around the globe. Thus, the "Dark Ages" were no more "dark" than subsequent or earlier eras. Moreover, the events and processes referred to as the scientific revolution were not the kind of singular striking event that the term "revolution" suggests, but rather one thread in long and slow processes of social formation. This chapter shows how the Voyages of Discovery provided another significant thread intertwined causally in those processes of social formation of modern sciences and technologies as well as of modern political relations, economic relations, and so on. Indeed, the Voyages of Discovery and the emergence of modern sciences were in part constitutive of those modern economic and political relations. And, of course, what was "discovered" was not primitive, timeless peoples with no histories—including no scientific and technological traditions.[28] Instead, the Europeans encountered complex and sophisticated cultures with highly effective scientific and technological systems, significant portions of which were "borrowed" into the European traditions, and other significant portions of which were destroyed either intentionally or inadvertently. Thus, postcolonial science and technology studies provide new geographic maps and historical chronologies within which to locate scientific and technological traditions, both European and non-European.

These expanded geographical maps and historical chronologies bring postcolonial and post-Kuhnian science and technology studies into convergence in other ways that highlight the gaps between the conventional epistemologies and philosophies of sciences and the ones we need to more accurately and usefully theorize how knowledge has been, is, and can be produced.[29] Both schools of post–World War II science and technology studies demonstrate that there is no one scientific method, no monolithic "science," and no single style of good scientific reasoning since both different eras of European sciences and the sciences of other cultures have used different methods and styles of reasoning to detect and explain systematic patterns of nature's regularities. Both show how scientific technologies and research practices have epistemological consequences. The postcolonial accounts examined in this chapter argue that the Voyages of Discovery and the practices of colonialism were themselves in effect research technologies with epistemic consequences no less significant than the laboratory practices and equipment of northern high sciences. Moreover, postcolonial studies join post-Kuhnian studies in insisting that it is scientific communities, not individual scientists, that provide the most revealing objects of analysis for charting the histories, cultures, and practices of scientific and technological change. And both argue that "scientific community" must refer to a much larger cast of social actors than the great men and their colleagues and scientific ancestors that constituted such communities in the conventional accounts.

The southern science and technology studies provide more evidence for the post-Kuhnian claims that every element of scientific and technological change in modern science has been contested, and, with but rare exceptions, that such contests have subsequently been hidden in the accounts told by the winners of such contests. They provide yet more evidence for the post-Kuhnian claim that it is not possible in principle to specify what constitutes "real science" in a way that makes the emergence, persistence, or decline of scientific claims and practices purely internal—or, for that matter, purely external. Thus, for postcolonial science and technology studies too, the epistemic status of sciences and technologies are always socially negotiated.

The postcolonial accounts use the kind of standpoint epistemology articulated most fully recently in the northern feminist accounts. Such epistemologies direct researchers to start from outside the dominant conceptual frameworks in order to identify key features of those frameworks that are difficult to detect from within them. Like the feminist standpoint epistemology and much of the post-Kuhnian science and technology studies, the postcolonial accounts acknowledge the socio-logical or historical relativism of scientific and technological claims and of their own accounts. Nevertheless, they, too, recognize that this does not commit them to cognitive relativism since they, too, hold that not all claims that sciences or science studies make are equally accurate.

Thus, the historical and philosophic issues raised by the kinds of postcolonial accounts examined in this chapter and the last converge in important respects with central directions in post-Kuhnian science and technology studies. The philosophic issues will be pursued further in later chapters.

The next chapter explores further what it means to think of modern sciences and technologies, like those of other cultures, as not just moti-vated, "provisioned," or influenced by local resources, but rather as deeply and completely constituted by them. Of course modern sciences are much more powerful and accurate in many respects than the sci-ences and technologies of other cultures. So in what senses would it nevertheless be accurate and useful to think of them as also "ethno-sciences"?

4

Cultures as Toolboxes for Sciences and Technologies

1. A Europology of modern sciences? Postcolonial histories and studies of contemporary projects have shown that in important respects modern sciences and technologies, no less than other cultures' traditions of systematic knowledge, are local knowledge systems. They are not alone in finding that a focus on the local provides more accurate histories and sociologies of modern sciences and technologies, for they are joined here in their own ways by post-Kuhnian and northern feminist accounts. Of course much of modern science is far more accurate at predicting many more of the regularities of nature's order than are many of the claims of other cultures' knowledge systems. However, as postcolonial and feminist studies have shown, modern sciences also produce patterns of systematic ignorance, and other scientific and technological traditions are more accurate at many of their own projects than are modern sciences at those same projects.[1]

Postcolonial science studies charted relationships between the advance of European scientific and technological representations of nature and of science on the one hand, and economic, social, and political projects of European expansion from 1492 (and earlier) on through late-twentieth-century so-called development policies on the other hand. It turned out that these two great processes marking "modernity"—the emergence of modern sciences in Europe and European expansion—provided crucial cognitive resources for each other as well as economic and political ones. In contrast to the older diffusionist models of the largely one-way spread of scientific and technological ideas and accomplishments from European to other cultures, such accounts have revealed many-directional interchanges between different cultures around the world, and how significant parts of the growth of modern sciences in Europe can be attributed to the successes of predatory expansionist policies.

Thus, postcolonial, post-Kuhnian, and northern feminist schools of science studies have all in different ways challenged the older, "internalist" philosophies of science that through their "rational reconstructions" of idealized scientific processes attributed the great successes of modern European sciences solely to science's internal epistemological features—plus, of course, the way nature is organized. Chief among the candidates proposed for such internal epistemological features of modern sciences have been the institutionalization of a critical attitude toward conventional beliefs, a distinctive method, uniquely high standards of objectivity, a distinctive rationality, a distinctive metaphysics that distinguished primary from secondary qualities, the shift from an organicist to a mechanistic model of nature, and the reliance on mathematics. Valuable as such features of modern sciences have been, such internalist accounts ignore the crucial resources that the sciences received from the "free ride" that European expansion provided.

We have learned in recent decades that Europeans produced an "orientology" of other cultures' beliefs and practices, including their accounts of the natural world. In the postcolonial science studies, a kind of Europology emerges—a delineation of distinctive characteristics of European culture and practices, including beliefs about nature and about sciences and technologies.[2] Post-Kuhnian northern science studies focussed on features of modern sciences and technologies that were local within Europe—late-eighteenth-century French, or early-twentieth-century U.S., or Protestant, or linked to the rise of the bourgeoisie, or of "polite society." Northern feminist accounts focussed on historically shifting androcentric social and scientific projects within European and North American cultures. Southern science studies focus on a third kind of localness of modern sciences—the features they share with European culture(s) that contrast with those in other parts of the world, and, in some cases, that became especially visible through their relations with other cultures.

The next section identifies four distinctively European features of modern sciences and technologies that have been described in postcolonial science studies. Section 3 then steps back to reflect on inevitable constraints on the cultural neutrality of any society's sciences and technologies. The argument here will be that such "constraints" also often provide crucial resources for the production of knowledge. That is, though culture in some respects functions as a prison house for the growth of knowledge, in other respects culture is its toolbox that enables valuable though always local knowledge about natural and social worlds.

2. European cultural elements of modern sciences.[3] Christian laws of nature. Let us begin with an argument by Joseph Needham, who could be regarded as one of the early contributors to postcolonial science studies.

He pointed out that the European conception of laws of nature drew on both Judeo-Christian religious beliefs and the increasing familiarity in early modern Europe with centralized royal authority, with royal abso-lutism. The idea that the universe was a "great empire, ruled by a divine Logos" was never comprehensible at any time within the long and culturally varying history of Chinese science that was the object of his studies.[4] A common thread in the diverse Chinese traditions was that nature is self-governed, a web of relationships without a weaver, in which humans intervened at their own peril.

> Universal harmony comes about not by the celestial fiat of some King of Kings, but by the spontaneous co-operation of all beings in the universe brought about by their following the internal necessities of their own natures. . . . [A]ll entities at all levels behave in accordance with their position in the greater patterns (organisms) of which they are parts. (323)

Compared to Renaissance science, the Chinese conception of nature was problematic, blocking interest in discovering "precisely formulated abstract laws ordained from the beginning by a celestial lawgiver for non-human nature."

> There was no confidence that the code of Nature's laws could be unveiled and read, because there was no assurance that a divine being, even more rational than ourselves, had ever formulated such a code capable of being read. (327)

Of course such notions of "command and duty in the 'Laws' of Nature" have disappeared from modern science, replaced by the notion of statis-tical regularities that describe rather than prescribe nature's order—in a sense, a return, Needham comments, to the Taoist perspective.[5] And yet other residues of the earlier conception remain in western sciences. For example, Evelyn Fox Keller has pointed to the regressive politics of the language of "laws," and to the positive political implications of concep-tualizing nature simply as ordered rather than as law-governed.

> [L]aws of nature, like laws of the state, are historically imposed from above and obeyed from below. . . . The concept of order, wider than law and free from its coercive, hierarchical, and centralizing implications has the po-tential to expand our conception of science. Order is a category compris-ing patterns of organization that can be spontaneous, self-generated, or externally imposed.[6]

European sciences advanced because of the constitution of their projects through these Christian and absolute monarchical assumptions, values, and interests. However, as Needham pointed out, the very same Chris-tian culture retarded European astronomy relative to that of the Chi-nese, for the latter was not burdened with the Christian notion that the heavens consisted of crystal spheres. Needham's discussion is particularly interesting since he is showing how local values, interests, and discursive resources—metaphors and models of nature—advanced modern science in some respects and retarded it in others. Moreover, it is *religious* cultural elements that had this powerful positive effect on the growth of modern science in Europe, though it is common to assume that modern science can only conflict with religion. Needham emphasizes how Christian elements served both as a prison house and a toolbox for the growth of scientific knowledge.

European expansion: Creating patterns of knowledge and ignorance. Earlier chapters explored the mutually powerful effects of European expansion and the emergence and growth of modern science in Europe. One distinctively northern feature of modern sciences that becomes visible from the perspective of this kind of account is the selection of modern sciences' "problematics." Just which aspects of nature European sciences describe and explain, and how they are described and explained, have been selected in part by the purposes of European expansion. Of course these are not the only purposes shaping these sciences, but they are significant ones. The problems that have gotten to count as scientific ones in the modern North disproportionately are ones that expansionist Europe needed solved. One historian points out that during the British occupation of India, in effect "India was added as a laboratory to the edifice of modern science" (as noted in an earlier chapter).[7] We can generalize the point: the world was added as a laboratory to modern science in Europe through European expansion, and continues to function in this fashion today through the science and technology components of so-called development that are controlled by the cultures of the north.

It is not that everything done by these cultures is done for exploitative reasons or has such effects in the South. Rather, the claim is that the projects that northerners are willing to sponsor and fund tend to be only those conceptualized by those in the North who get to participate in making development policy and by their allies elsewhere. The majority of peoples who bear the consequences of the science and technology decisions made through such processes do not have a proportionate share in making them—to adapt to the world of sciences and technologies one formulation of a fundamental democratic ethic. The picture of nature produced by solving the problems identified by an expansionist North, or by northern conceptions of development, ignores or hides those aspects of nature that northerners assume are irrelevant to success at such projects. Both systematic knowledge and systematic ignorance are produced by northern sciences' distinctive "locations in nature" that were themselves in part created through European expansion, and by the kinds of interests that European cultures and their sciences had in those parts of nature.

As we saw in an earlier chapter, postcolonial science and technology studies pointed out that modern sciences answered questions about how to improve European land and sea travel; mine newly needed ores; identify the economically useful minerals, plants, and animals of other parts of the world; manufacture and farm for the benefit of Europeans living in Europe, the Americas, Africa, and India; improve their health and occasionally that of the workers who produced profit for them; protect settlers in the colonies from settlers of other nationalities; gain access to the labor of the indigenous residents; and do all this to benefit only local European citizens—for instance, the Spanish versus the Portuguese, French, or British. But they have not been concerned with explaining how the consequences of interventions in nature for the benefit of Europeans would change the natural resources available to others, or what the other social, psychic, environmental, economic, and political costs of such interventions might be. They have not been concerned to explain how to eradicate diseases or other health-threatening conditions that do not much affect peoples of European descent, and, especially, the already advantaged within this group. They have not been concerned to explain how to effectively use sustainable energy sources, to maintain sustainable food supplies for southerners, or, until recently, how to sustain fragile environments anywhere on the globe or surrounding it. Even physics, supposedly the most value neutral of sciences, is far more shaped by its pursuit of militarily useful knowledge than is generally recognized.[8]

Moreover, as earlier chapters described, European sciences borrowed elements from other scientific traditions. The record of such borrowings charts the distinctive trajectory of the Voyages of Discovery and their later descendants, including today's development projects. And more than mere borrowing was made possible through European expansion; such elements were not merely taken into European sciences as isolated elements, but combined with others and transformed into new kinds of scientific and technological knowledge. Brockway's account of the role of Kew Gardens in the development of plantation agriculture is just one report of such a process.[9]

Thus, the distinctive patterns of knowledge and ignorance characteristic of modern sciences are in significant part products of European expansion. Expansionist projects gave European sciences distinctive opportunities to chart nature's regularities and distinctive interests in which regularities they would encounter and which of those they would chart. The cognitive successes and also the failures and gaps of modern sciences are importantly due to, and in their representations of nature bear the distinctive historical marks of, European expansion.

Northern distribution and accounting practices. A third element of a Europology of northern sciences is the distinctive pattern of the distribution of their consequences—who gets which consequences of scientific and technological change—and the way northern sciences account for these distribution patterns. Postcolonial accounts argue that the accounting practices mask the actual distribution of sciences' benefits and costs.

The benefits of modern scientific and technological change are disproportionately distributed to elites in the North and their allies in the South, and the costs disproportionately to everyone else. Whether it is sciences intended to improve the military, agriculture, manufacturing, health, or even the environment, or "pure sciences" doing "fundamental" or "basic" research, the expanded opportunities sciences make possible have been distributed predominantly to already privileged people of European descent, and the costs to the already poorest, racial and ethnic minorities, women, and Third World peoples.[10]

The causes of this distribution are not mysterious or unforeseen. For one, the postcolonial accounts point out that it is not "man" whom sciences enable to make better use of nature's resources, but only those already advantageously positioned in social hierarchies. It is such groups that already own and control both nature, in the form of land with its forests, water, plants, animals, and minerals, and the means to extract and process such resources. Moreover, these people are the ones who are in a position to decide "what to produce, how to produce it, what resources to use up to produce, and what technology to use."

> We thus have this spectacle, on the one hand, of the powerful development of technological capacity, so that the basic and human needs of every human being could be met if there were an appropriate arrangement of social and production systems; and, on the other hand, of more than half the world's population (and something like two-thirds the Third World's people) living in conditions where their basic and human needs are not met.[11]

Such critics thus confirm how the opportunity to construct the most powerful representations of nature—to produce knowledge that more accurately predicts more of nature's regularities—is available only to already advantaged groups and is chosen in ways that will increase the advantage such groups already have.

However, they go on to point out that it is through science's accounting practices that this distribution is kept invisible to most of those who benefit from modern sciences and to many who do not. All consequences of sciences and technologies that are not planned or intended are externalized as "not science."[12] Such an accounting need not even be intended; critics argue that such an "internalization of profits and externalization of costs is the normal consequence when nature is treated as if its individual components were isolated and unrelated."[13]

Thus, they argue that at its cognitive core, distinctively European interests, discursive resources, and ways of producing knowledge constitute
modern European sciences in ways that have distinctive political dimensions and consequences.

Value-neutrality is not value neutral. A fourth distinctively European
component of the cognitive core of modern science that has been identified in many of the postcolonial science studies is the claim to, and
valuing of, cultural neutrality. Even if it were the case, impossible
though it be, that modern sciences could bear no such cultural fingerprints as the kinds marked above, their value-neutrality would itself
mark them as distinctively European. Of course this is paradoxical:
"if it's value free, then it's not value free." The point is that trying to
maximize cultural neutrality, as well as claiming it, expresses a culturally specific value. Most cultures do not value neutrality; they value
their own Confucian, or indigenous American, or Islamic or Maori or,
for that matter, Judaic or Christian values. So one that does value and
maximize apparent neutrality is easily identifiable.

Moreover, the claim to neutrality is characteristic of the administrators of modern cultures that are organized by principles of scientific
rationality, as feminist and class theorists have argued.[14] Abstractness
and formality express distinctive cultural features, not the absence of all
culture. Thus, when modern science is introduced into other cultures, it
is experienced as a rude and brutal cultural intrusion because of this
feature, too, point out the postcolonial accounts. Claims for modern
sciences' (value-neutral, internally achieved) universality and objectivity are "a politics of disvaluing local concerns and knowledge and legitimating 'outside experts.'"[15]

There are other distinctively European features of the cognitive cores
of modern sciences and technologies that emerge if one starts thinking
about them from outside their own conceptual frameworks, as postcolonial science studies have done. Let us now step back from these
particular accounts—and many more are provided in the post-Kuhnian
and feminist accounts—to examine in greater detail how it is that any
and every culture's scientific and technological traditions will bear
traces of the local resources that constituted them.

*3. Cultures are "toolboxes" as well as "prison houses" for sciences and
technologies.* We can identify several distinctive kinds of cultural elements that come to constitute the cognitive cores of any systematic
knowledge about the natural world under any social and political conditions. The four examined here have been selected because extensive
literatures familiar to at least some readers are available in each case to
enable us to understand just how it is that, in some respects, cultural
specificity could, should, and does have positive effects on the growth
of knowledge. (There may well be other useful ways to divide up these

"causes" of the necessary cultural features of scientific traditions.) We shall consider cultures' locations in heterogeneous natural regularities, which enables us to draw on familiar approaches from biology and geography; culturally different interests that lead to valuably different questions about such natural environments, a familiar approach in sociology and political economy; cultures' distinctive discursive traditions, that enable them to "see" nature in new ways, an approach familiar from interpretive social sciences and cultural studies; and, last, culturally distinctive ways of organizing the production of knowledge, an approach familiar in anthropology and the sociology of knowledge.

Locating cultures in heterogeneous nature's order. Nature is not uniformly organized. Individual societies do not have access to observation of the entire diversity of nature's features—not even when expansionist projects or other forms of world travelling take them far from home. The very survival of societies depends upon their ability to interact effectively with their own, local share of nature's regularities—even when that local share is spread through their travel from Genoa to the Caribbean or from Cape Canaveral to the moon and beyond. Some cultures daily interact with high altitudes and others not; some with mountainous terrains, deserts, oceanic islands, rain forests, or rivers; some interact with extremely cold and others with extremely hot climates; some with one range of diseases and health hazards, and others with quite a different range. People in each culture need to be able to protect themselves from the natural patterns peculiar to those particular climates, land formations, plants, animals, and diseases that surround them as they move about, and to figure out how to gain access to the potential resources for food, clothing, shelter, travel, exchange, and other needs that their part of nature offers. Moreover, biological differences—dark skins or light, immunity or not to malaria, and so on—create different interactions with surrounding environments.[16]

History is full of accounts of the extinction of groups of humans who tried to settle into an environment that turned out to be unremittingly threatening to their biological makeup, or from which they had not yet learned effective ways to gain resources for survival. For example, European attempts to settle in the Americas prior to 1492 were not successful because the Europeans were unable either to establish the environmental "little Europes" that would enable them to cultivate familiar food, shelter, and other resources on this new continent or to develop the local resources of the continent into functional food and shelter. Likewise, the European settlement of tropical Africa was delayed until the late nineteenth century because Europeans could not survive the diseases to which indigenous Africans had developed immunities.[17] Moreover, the Europeans introduced new diseases into the Americas that quickly decimated the indigenous peoples. As one account puts the point, America was not conquered; it was infected.[18]

Nor have the great successes of modern medicine reduced the probability of disequilibriums caused by infectious diseases continuing regularly to emerge in humans and their environments. Nature is not static; it changes in unpredictable ways. Microbes continually evolve and try to eat us. AIDS, tuberculosis, *E. coli*, malaria and a whole new group of "flesh-eating" diseases generate a continuing juggernaut that epidemiologists can hope, at most, to slow down. Moreover, diseases obviously are spread through many social practices such as disrupted environments, global travel, tolerating poverty, generating population explosions, centralizing medical practice in hospitals, or a collapse of public health institutions.

Such situations clarify that a culture's "local environment" is not geographically fixed or historically static. It is useful to think of it as whatever the route through nature's heterogeneous patterns of regularities along which the culture's particular way of life takes its members—a kind of historically and geographically distinctive environmental tunnel. Thus, Spain-to-the-Caribbean should be considered such a local environment within which humans travelled in both directions. More accurately, it is not individual humans but "biotic packages," as historian Alfred Crosby puts the point, that make such journeys. Humans, their dogs and sheep, the germs, fleas, lice, ticks, and other animals that hitched rides on all three species, the ships' rats, the rats' fleas, the fleas' parasites, the diverse plant seeds that happened to travel inside and on the outside of all of these animals, the plants the travellers intentionally brought with them and the microbes living in the plants—this biotic package travelled from Spain to the New World. Back to Spain travelled similar packages of humans and their biotic companions and dependents. In such ways, local environments are constantly expanding and contracting. Nor, as indicated, are such environments limited to this planet, for Cape Canaveral-to-the-moon is another such historical/ geographic environmental tunnel. Such environments are always historically changing and geographically expanding and contracting, and they lead to humans' exposure and biological or cultural adaptation (or not) to diverse regularities of nature and their underlying causal tendencies.

With lots of people moving in lots of directions now one might think that there are no longer distinctive local environments—that the world is just one big environment for us all in which the ebola virus can with frightening rapidity become part of the local environment of Washington, D.C., through the importation of African monkeys into a U.S. federal research lab, and the air inside any major airport becomes the shared environment of people from every nation in the world. However, even with the shrinking of the world, with greatly increased travel by tourists, executives of multinational corporations and international agencies, armies, refugees, and so on, environments still remain usefully thought of as local. For one, no one can be everywhere at once, so everyone still lives in their own local environment no matter how extended through space it may be. Moreover, since so-called natural patterns change historically as human activities change, patterns of local environmental change continually are formed and reformed in unforeseen and often unforeseeable ways.

Theories that appear plausible in one environment may not appear so in another. For example, cultures living on the edges of continental plates might well find the geology of plate techtonics more plausible (and "interesting"!) than do cultures with little experience of earthquakes, volcanoes, and other phenomena characteristic of plate-juncture environments. The latter may have developed geological theories based on their experiences of the vertical movement of the earth visible in mountain ranges—theories that appear implausible and/or uninteresting to the plate-juncture dwellers. Theories that identify physical causes of premenstrual syndrome, chronic stress syndrome, or chronic fatigue syndrome may appear more plausible to people in whose bodies these phenomena occur, while others may find psychosomatic explanations of these symptoms more reasonable. Physiological theories of the causes of obesity or of alcoholism may also appear more reasonable to those in whose bodies such phenomena occur, while "bad habits" may be the favored forms of explanation for those not beset by such conditions—or vice versa.

Note that the claim here is not that knowledge based on some set of local interactions with nature is always more accurate; obviously very often it is not. The fact that knowledge is local is no guarantee that it is the most accurate, let alone the most comprehensive. Rather, the claim is that cultures' different locations in heterogeneous nature expose them to different regularities of nature or to what appear to them to be regularities of nature, and that exposure to such local environments can be a valuable resource for advancing knowledge. Different peoples are repositories for historically developed and continually refined beliefs about different parts of nature.

A skeptic about the importance of the argument here might well insist that modern science, in contrast to the mere local knowledge systems that have developed in these different natural niches, can in principle predict and explain all of nature's diverse features. Thus, it is in a position to fit together like a jigsaw puzzle descriptions and explanations of different parts of the natural world. At the end of the next section we return to identify the limitations of this common assumption.

Different social interests. Even in "the same" environment different cultures have different interests and desires. These lead cultures to pose distinctive questions about "the same" part of the natural world.[19] Sciences' problematics are shaped by their supporters, sponsors, and funders and, more generally, by what is interesting to those groups willing and able to have their concerns conceptualized as ones for systematic empirical research. Bordering the Atlantic Ocean, one group will want to fish it, another to use it as a coastal highway for local trading, a third to use it for trans-Atlantic emigration or trading slaves for sugar, a fourth to desalinize it for drinking water, a fifth to use it as a refuse dump, a sixth to use it as a military highway, and a seventh to mine the minerals and oil beneath its floor. These differing interests have created culturally distinctive patterns of knowledge (and ignorance) about this part of nature's regularities and their underlying causal tendencies.

Many histories of science familiar to some readers provide examples of how culturally different interests have shaped distinctive patterns of scientific knowledge and ignorance. For example, there is Donna Haraway's account of how different national preoccupations generated different knowledge in different countries' primatology studies. Thus, Anglo-Americans have been preoccupied with sex and power relations among primates, but Japanese primatologists have not. The latter have instead learned how cultural differences are developed among ethnically similar groups.[20] Historian Robert Proctor has shown how the differing interests and power of medical and environmental research lobbies have shaped both what we know and what we do not know about the causes of cancer. In spite of the well-known fact that patterns of cancer incidence are highly correlated to levels of carcinogens in the environment, environmental research has not been highly funded. Instead, the majority of research has examined variation in individuals' susceptibility to carcinogens due to their genetic makeup or their "lifestyles." Clearly, medical research lobbies are far more powerful and effective at getting funds than are environmental research lobbies. However, behind this fact one can observe the reality that environmental research into carcinogens challenges the current and past practices of three of the most powerful kinds of institutions in the world today: militaries, multinationals, and governments. The search for the causes of cancer to be found in individuals, their lifestyles, and genetic inheritances, does not have this effect. These three kinds of powerful organizations have kept us ignorant in many other ways about how environments are destroyed and what must be done to repair them.[21] Even contemporary physics—some would say, especially contemporary physics—has not been immune to direction by special interests. Paul Forman has documented how national security interests came to direct the agenda of physics in the United States before and during World War II.[22]

The point in this section is that what gets to count as interesting scientific questions depends in part on the interests different cultures and subcultures have in learning about those of nature's regularities and their underlying causal determinants to which they are exposed. In

"the same" environment, different groups such as medical and environmental researchers will be interested in different questions. Thus, it is
illuminating to read the history of modern science as a history of local
interests. There is little inevitable about the particular pattern of knowledge that we think of as due to the inevitable forward march of scientific
research. Of course we have only the history we have managed to
accumulate, so we can only imagine what patterns of knowledge could
have been produced by cultures with interests different from those that
directed the route of European sciences. However, in focussing on new
directions in northern sciences that could have been more highly funded
and institutionalized earlier (environmental sciences, epidemiology),
on the distinctiveness of patterns of knowledge created through European expansion, and on the patterns of knowledge characteristic of
other cultures, we can find resources for such thought experiments.

There is nothing controversial about observing that social interests
shape what gets to count as interesting scientific questions. What is
controversial here is to claim that science, real science, includes the
choice of scientific problems; to point out that the cognitive content of
science is shaped by and has its characteristic patterns of knowledge and
ignorance precisely because of problem choices; and to argue that
different interests produce not just different pieces of the puzzle of
nature's regularities and their underlying causal tendencies, but fundamentally incompatible knowledge claims. Referring to feminist criticisms of sciences' inattention to women's health and medical concerns,
Evelyn Fox Keller characterizes the conservative view this way: "[T]his
kind of criticism does not touch our conception of what science is, nor
our confidence in the neutrality of science. It may be true that in some
areas we have ignored certain problems, but our definition of science
does not include the choice of problem—that, we can readily agree, has
always been influenced by social forces."[23] However, this older restrictive definition of science is no longer the most authoritative one, for it
is now widely understood that the content of scientific knowledge is
importantly shaped by interests in knowledge and interests in ignorance, as Keller among others has pointed out. Even the U.S. National
Academy of Sciences—no one's idea of a radical science enterprise—has
made clear that the older definition must be expanded. The notion of
scientific methods, they say, can reasonably be broadened to "also
include the decisions scientists make about which problems to pursue
or when to conclude an investigation. . . . Taken together, these methods
constitute the craft of science."[24] The craft of doing science is to include
defensible decisions about which problems to pursue and when to
regard an investigation as completed. What appeared in conventional
accounts of science as the apparently innocuous pursuit of truth wherever a scientist's interests might take him or her has become a far more

interesting origin of culturally distinctive patterns of systematic knowledge and systematic ignorance.

"One elephant and one true account of him"? "But don't these differing exposures to and interests in nature simply result in our ability to fill in more and more pieces of the potentially one, true representation of nature?" asks our skeptic. After all, shouldn't we think of the history of science as something like the story of the five blind men and the elephant in the proverbial account? Let's make it ten blind men—five cooks and five transportation managers—and an elephant, so that we can incorporate not only exposure to different parts of the elephant but also different interests in it. Our blind men are assigned two to the tail, two to the ear, two to the trunk, two to the body and two to a leg, and in each pair one is interested in cooking and one in transportation. Exposed to different parts of the elephant and with differing interests, won't they eventually be able to piece together a coherent story of "the real elephant"?

Not if the analyses of these three schools of post–World War II science studies are right—an issue to be pursued in greater detail in chapter 10. To preview the discussion there, in the first place John Dupre has pointed out how nature itself is disunified. The dream of a unity of nature that justified the positivist unity of science programs in the early twentieth century is only a delusion since nature itself is heterogeneously organized such that there can be no one uniquely valid representation of it: "... the disunity of science is not merely an unfortunate consequence of our limited computational or other cognitive capacities, but rather reflects accurately the underlying ontological complexity of the world, the disorder of things."[25] The underlying assumptions of the unity of nature thesis are at odds with the sciences' own findings. The metaphysics upon which the unity of science thesis depends is inconsistent with sciences' best results of research, Dupre shows.

In the second place, even if nature were unified, the representations of nature produced by differing interests could not all fit together, like pieces of a jigsaw. Insofar as distinctive interests lead to distinctive interventions in nature, the nature around us changes. We can no longer be interested in species that have disappeared in the ways we would have been while they still offered possibilities of resources for food, pharmacology, transportation, companionship, and so on. Can the environment always repair itself in response to toxicity and other forms of human-induced changes? The interventions to which particular, interested representations of nature lead both enable and limit the knowledge possible about nature.

Moreover, different interests lead to different ways of representing nature, and are themselves shaped by them—a point to which we shortly turn. Thus, we cannot really speak of a culture's interests as if these were objectively specifiable apart from the cultural presuppositions, metaphors, models, and narrative strategies that formed that culture's distinctive inheritance and that shaped interests in, for example, eating animals (and one of *that* kind, that other cultures prohibit eating). What a culture perceives as its natural environment and its interests are both mediated by the culturally local discourses available to it—for example, Christian or Confucian understandings of law and of nature's order, or perceptions of nature as organic or mechanistic.

Thus, systematic ignorance is always also produced along with systematic knowledge, and these patterns of ignorance are just as culturally distinctive as are a culture's patterns of knowledge. The unity of science theses are already in trouble before we even get to the features of sciences upon which most critics of these theses have focussed.

Discursive resources for representing and intervening in nature. It is with the multiplicity of discursive resources available to scientific cultures that critics of the "one true story" accounts of scientific knowledge have most been concerned. The role of metaphors, models, and narratives of nature in formulating research projects provides important examples of this rich category.[26] Postcolonial, feminist, and post-Kuhnian northern science studies have pointed out how such cultural elements insure that modern sciences have not been, are not, and could not be value free. The goal of absolute value freedom is counterproductive, such studies argue, because the advance of scientific knowledge requires just such culturally local discursive resources to direct its theories and subsequent interactions with nature. Thus, cultural presuppositions are not the unmitigated defect in the sciences that they have conventionally been thought to be.

According to these post–World War II analyses, scientific theories are underdetermined not just by any existing collection of existing evidence for them, but by any possible such collection of evidence. Scientific processes are not transparent; their culturally regional discursive resources are one of the kinds of cultural elements that conjoin to constitute the conceptual frameworks for descriptions and explanations of nature's order.[27] Thus, more than one scientific theory or model can be consistent with any given set of data. Moreover, each such theoretical representation can have more than one reasonable interpretation. Indeed, this looseness or slack in scientific explanation, far from being the unmitigated defect that it appears to be in older philosophies of science, turns out to be a major source of the growth of scientific knowledge. It is this feature that permits scientists to "see nature" in ever new ways that advance the increased accuracy and comprehensiveness of their claims. The cultural presuppositions, metaphors, models, and narrative structures constituting mechanistic views of nature give scientists one important set of guidelines for how and where to extend their theories and for where to revise them when confronted with surprising observations.[28] These cultural features of sciences' cognitive cores shape how people see and consequently intervene in nature.

Thus, different cultural elements provide crucial resources for advancing the growth of knowledge. Such elements can be borrowed by a science from other disciplines, directly from the surrounding culture, from adjacent cultures with which a science interacts, from older elements of a culture's own scientific tradition, and, no doubt, from other sources. Many thinkers have perceived the language-dependency of scientific accounts only in negative terms—as the "prison house of language." The post–World War II science studies agree that the metaphors, models, narratives, and other discursive features of the sciences can indeed constrain the growth of scientific knowledge. However, they also provide invaluable resources for description, prediction, explanation, and understanding of nature's regularities. Scientific language necessarily both enables and limits what a culture can know about nature's regularities and their underlying causal tendencies. It is a source of different sciences even within the history of modern science as the favored models of nature, evidence, good method, and the like have changed. The cultural features of scientific language enable cultures to draw on their own familiar metaphors and models to explore different aspects of nature's regularities; they take the standpoint of their cultural inheritances in order to "see nature" in distinctive ways—a feature of science that is also highly motivating to the production of knowledge about our environments.

Thus, in this third respect, also, local knowledge proves to be a resource for the collective growth of human knowledge about that natural world. Moreover, since these cultural elements are not detachable from some imagined culturally neutral representations to be found within or behind them, no hope is offered here, either, for imagining that different sciences can simply fit together culturally neutral images of nature, leaving behind the culturally local. While there are important respects in which organismic and mechanistic representations of nature can accommodate "the same" observational data about, for example, the movements of the stars and planets, historians show us how each also generates information about the natural world that is not containable within the other set of discursive resources and that in some respects conflict.

Culturally local ways of organizing scientific work. Last, the cognitive content of sciences is shaped by culturally different forms of the social organization of scientific research. Scientific research is social labor, carried out in culturally distinctive kinds of organizations—laboratories located in industries, universities, physicians' offices, federal institutes or computer-connected collections of such sites; field stations, farms, collecting and observing expeditions, conferences, learned societies, journals, hospitals, routine visits to healers with culturally diverse credentials, and so on. What we can know of nature's regularities and their underlying causal tendencies depends in part on how communities for systematic observation and explanation of nature's regularities are organized.

Most obviously, what we know depends upon which kind of disciplinary organization of research is favored. We have already noted how our knowledge of the causes of cancer will vary depending upon whether the interests of medical or environmental researchers are favored. But these two fields not only have different interests, but different ways of organizing the production of scientific knowledge; medical researchers organize in one way their search for the causes of cancer, and environmental researchers do so in another way. What we will and will not know about the regularities of cancer and their underlying causal tendencies is shaped by how cancer research is organized.

In a second kind of case, cultures come to develop skills and abilities that enable them to understand the natural world in distinctive ways. In her primatology studies, Haraway discusses that Japanese women primatologists routinely learn to identify individually huge numbers— hundreds—of the primates they study. This is a feat that Japanese men find remarkable, she reports. This skill enables these women to learn far more about the complex social relations in these species.[29] In another case, Michael Baxandall reports that because merchants and their buyers in the early Renaissance were expert at the practice of "gauging" the volume of barrels and other containers of loose goods, audiences for the elaborate perspective paintings of the period—with all of their columns, vaults, and numerous walled spaces—could see the spaces so contained in ways that we no longer can today. Pleasure and skill at perspective "seeing" made possible the development of important parts of early modern sciences.[30] Steven Shapin and Simon Schaffer show how the scientific experiment was "invented" through careful staging of demonstrations of the air pump to suitable observers.[31] Bruno Latour argues that Louis Pasteur had to turn physicians into scientists in order to demonstrate the validity of his hypotheses.[32]

Moreover, the accounts of how European expansion and the growth of modern science required each other can be read as arguing that modern sciences were constituted in part by the organization of European expansion. As we saw in an earlier chapter, European expansion required better cartography, astronomy, oceanography, shipbuilding, geology, and mining; better understanding of the plants, animals, diseases, and peoples of other parts of the world; and diverse other knowledge of natural and social worlds. Further, those sciences took their focuses, borrowed crucial elements, and were able to expand the domain of their validity through the resources that European expansion made possible. European expansion was a form of organization of the production of modern sciences' knowledge. Cultures that could not benefit from this kind of collecting and integrating of knowledge from many diverse parts of nature and cultural traditions lost knowledge (not to mention, often, their lives and cultures) because of European expansion.

As yet another kind of example, Sharon Traweek's comparative study of Japanese and U.S. high-energy physics shows how cultural models of the organization of work and ideal career trajectories, on the one hand, and nationally established relationships between equipment manufacturers and physicists on the other hand, converge to generate the pursuit of at least marginally different scientific problems in Japan and the United States. In the United States, people rise through professional careers primarily as individuals. They join a research team for a few years and then advance further by moving to another, sometimes as the original one breaks apart. In Japan, workers join a workplace team with which they may stay for the rest of their working lives. This cultural difference in the organization of work would in itself tend to make Japanese experimentalists willing to choose longer-lasting research projects than U.S. experimentalists would favor. However, this possibility is enhanced by the differing relationships to their technologies characteristic of the two national research communities. U.S. experimentalists in the high energy physics community constantly tinker with their equipment, adjusting it to make it perform better or just differently. In Japan, the detectors are made by outside manufacturers, and experimentalists are not permitted to tinker with the machines. Thus, they must work with the apparatus they have ordered until they can justify the huge investment of a new one. In both ways, the Japanese model encourages longer-lasting research projects than does the U.S. model. Do they gain different knowledge about nature? Traweek does not say so, but this example draws our attention to how the organization of scientific research is shaped by cultural features organized outside scientific communities.[33]

Ideal or characteristic lengths of research projects can also be culturally affected in other ways, such as by how closely the funding of research is tied to electoral politics. In societies where there is a very close relationship—where funding agencies have little autonomy from the party in power—short-term scientific projects tend to be the main ones. This is because no politician wants to waste money supporting research the results of which will not be available until after he or she is out of office and able to benefit from its sponsorship.

It is useful to recollect that Thomas Kuhn had argued that modern science's successes were due to the distinctive way scientific communi-

ties were organized in the modern West.[34] There is nothing controversial about arguing that in big and little ways, how the production of
scientific knowledge is organized will shape the representations of
nature sciences produce. What is controversial is to see that this feature,
too, insures that all knowledge, no matter how universally valid it is
regarded to be, is local knowledge. Consequently, it appears that there
can be many culturally distinctive and yet scientifically valuable ways
of organizing the production of knowledge about the natural world.

4. *Conclusion.* This chapter has identified four ways in which the
conceptual framework and content of any and all scientific and technological traditions, including that of modern European sciences, will be
culturally local. Modern sciences appear as distinctively European from
the perspective of cultures that have not shared European geographical
locations, interests, discursive traditions, or ways of organizing the
production of knowledge. However, modern sciences' features are distinctively European in another, fifth, way that crosscuts or permeates
the other four. In the first four cases, European sciences are constituted
and maintained through the types of cultural resources of which any and
every culture has its own unique repository. In the fifth case, European
sciences have been positioned in a distinctive place in global power
relations that is not shared with other cultures; Europe's domain of the
"local" expanded in a fashion that was at least partially predatory of
other cultures' resources, and the localness of such European features
was made to disappear. Until recently, the European worldview seemed
the only reasonable one to Europeans and their admirers.

This account demonstrates that the invisibility to the "natives" of
widespread but local features of a conceptual framework cannot provide
evidence that such features do not exist, but only that no critical external
perspective has been available to reveal them. Their invisibility is a
sign of their hegemony. Thus, one can never assume that any particular
set of knowledge claims has escaped the local, for there will always be
other cultures and practices from which will become visible the localness
of such claims.

One can see here that the standpoint of the outsider (the first four
forms of the local) and of the exploited (the fifth) generating Europology
provide powerful resources for enabling sciences more accurately to
understand their own processes.[35] The following chapters turn to look
at how standpoints on nature can also be gendered and at the kinds
of feminist science studies that can emerge when such analyses are
located on the historical and multicultural maps of European and other
science traditions that we have been exploring in this and the preceding
chapters.

5

Postcolonial Feminist Science Studies: Resources, Challenges, Dialogues

1. Should feminist science and technology studies be culturally diverse?[1] Feminist theory is not now and has never been a monolithic whole. Indeed, a powerful stimulant to its growth and increasing usefulness has been the continual expansion of the historical, political, and cultural locations from which it has spoken. Different political philosophies have guided the analyses and forms of activism characteristic of liberal, Marxist, radical, and socialist feminisms—to name four of the feminisms that have most shaped public agenda projects in Europe and North America since the eighteenth century. Multicultural and global feminisms, which have analytic projects that are partially distinct, are two recently emerging and important feminisms designed to direct public policy. They provide resources for thinking about gender relations within multicultural societies and a multicultural world, and within global political, economic, and social relations as these are understood from postcolonial standpoints.[2] (I shall subsequently refer to them collectively as postcolonial feminisms.)

These concerns have not been central in constructing the conceptual frameworks of the liberal, radical and socialist feminisms within which contemporary northern feminist science and technology studies, including philosophies of science and technology, have been developed. This is not to say that postcolonial issues have been totally absent from these older theoretical approaches, but only that they have not been centered in them. Thus, the question here is whether and how the distinctive concerns of postcolonial feminisms call for distinctive approaches to science and technology issues.

This may seem an odd question to readers new to feminist theory or new to science studies. Those new to feminism often think that women, and their feminism, should speak with one coherent voice. Similarly, those new (or resistant!) to post–World War II science studies often think that we should have *a*, and preferably *the*, theory of science. We should not have many of them, each distinctive to different political theories, let alone to historical, cultural locations. Multiplicity is taken to be a sign of error from these conventional perspectives; or, at least, acceptance or appreciation of it is taken to reflect a damaging seduction by soft-minded relativists. Moreover, in the case of feminism, the "unitarians" often presume that the reason there are so many feminisms is that women are just inherently uncooperative with each other or unable to arrive at the kind of single approach that these critics imagine to be the only rational and effective one. If women were fully rational, they would join in support of one, true account of women's and men's conditions and what to do about them. However, as earlier chapters have already shown in some detail, just as modern and other sciences are all local knowledge systems, some of which are nevertheless immensely more powerful than others, so too are science studies, including feminist ones. There should be many feminist approaches to sciences and technologies because both science and technology studies and feminisms are inevitably and usefully multiple. This and the following two chapters explore ways in which gender issues have been and could be located in both postcolonial and northern sciences and technologies as well as science and technology studies.

It is easy to see why and how the social theories of these diverse feminisms lead to different concerns about scientific and technological change, and thus to different and sometimes conflicting focuses of their philosophies of science. Each of these social theories has started off from different groups of women's lives to think about gender relations, including those of scientific and technological institutions, their cultures and practices. Culturally different groups of women have tended to see their needs, hopes, and fears most accurately represented by different social theories. They have developed different feminist standpoints that see the world through the lens of conceptual frameworks that directly address the issues important to them. Two well-developed feminist theories of scientific and technological knowledge are feminist empiricism and feminist standpoint epistemologies.

Feminist empiricism has directly addressed issues important to liberal feminism. These empiricist philosophies of science have retained important elements of their origin within the conceptual framework of the eighteenth-century educated classes' concern with the primacy of reason, the importance of individual observation ("experience"), the formation and authority of distinctively qualified professional communities, separation between public and private, or unofficial, worlds of science institutions, and the importance of arriving at the one true story of nature's order. For them, women, too, whatever their obligations and activities in private life, can become rational observers and explainers of nature's order, no less than their brothers can. Science communities would benefit from their presence in greater numbers as women assist in the project of arriving at more accurate and less androcentric results of research. (Of course, feminist empiricisms are much richer than such a brief sketch can convey.) However, feminist science theorists working within this tradition have not simply "added women to empiricism." Though working within this particular social theory, they have at the same time innovatively challenged and transformed it in illuminating and valuable ways.[3]

It is those particular frameworks, however, the ones that shape the culture and practices of empiricist approaches to the sciences, that feminist empiricists have sought to transform. To say this is not to diminish their undoubted value or the brilliance and hard work it has taken for feminist empiricists to produce these important theories of knowledge that are being used by researchers and scholars in so many disciplines. Instead, it is to insist that feminist empiricism provides just one of the many valuable feminist approaches to science and technology studies that are now and will in the future become available to us. Feminist empiricism is not the one and only one, nor is it superior to the others in every way. This kind of clear positioning of feminist empiricism is necessary because, among the feminist science and technology theories, it is closest to the conventional philosophies of science and epistemologies that insist on epistemological "unitarianism." That is, the conventional philosophies insist that there exists one unified nature, and one true story about it, and that modern science—the only "real science"—can tell that one true story about nature's order. Residues of this unity of science thesis can be found most strongly in feminist empiricist philosophies of science and epistemologies.

Other philosophies of science indebted to socialist feminism retain elements of their partial origins in nineteenth-century Marxian analyses of the causal relations between what people actually do in everyday life ("labor," in the original Marxian account) and the kinds of conceptual schemes that they tend to find most plausible (what they can know), and thus a focus on how the hierarchial social relations that organize human labor lead to the assumption of conceptual frameworks that produce both systematic knowledge and systematic ignorance.[4] Beginning thought or a research project from the standpoint of women's lives can reveal aspects of the dominant institutions and their conceptual frameworks that are not visible if one starts from inside those institutions and the disciplines that rationalize them.

Feminist standpoint theories thus begin from the assumption that power and knowledge are inevitably linked, but that not all power or knowledge belongs to the powerful. In contrast, feminist empiricisms still hold to the ideal of desirable, and perhaps even possible, systems of knowledge that represent no particular set of social interests and desires, and thus that exist outside of societies' power relations. Their goal is to remove sexist and androcentric fingerprints on the results of scientific and technological research, while trying to avoid replacing them with any other such merely local characteristics, such as women's or feminist "fingerprints." They are after "pure science," not any other kind of purportedly better science. After all, if there were only one possible true story about nature's order, as empiricists hold, it could not be one shaped by merely contingent human interests and desires, such as women's or feminists'. Like feminist empiricism, feminist standpoint theory also has incisively challenged and usefully transformed key elements of the conceptual framework within which it has chosen to work. One such challenge/transformation found in some standpoint writings is the rejection of Enlightenment unity of science assumptions that shaped classical Marxian and socialist epistemology no less than it did contemporary empiricism. Standpoint theory will be discussed further in a later chapter.

Strains of these earlier feminist analyses appear in both multicultural and global concerns of postcolonial feminism that, drawing on earlier roots, became fully articulated in the 1980s. Multicultural feminisms focus on the culturally distinctive histories and practices shaping women's conditions, interests, and desires in different local, national, and transnational cultures, and the necessity for thinking from these distinctive histories in designing social change. African American women, Chicanas, Puertoriqueñas, Chinese-, Japanese-, Korean-American, and other Asian American women, Native American and Jewish American women—these as well as Italian Americans, Irish Americans, and women of other ethnicities and cultures bring different resources and different struggles to the feminisms they develop. Such feminisms must be in solidarity with each other, not unified into one, monolithic feminism that would inevitably lose sight of issues that differed, and sometimes conflicted, between such different groups of women. Multicultural feminisms have also focussed attention on the powerful resources of bi- and multicultural communities in creating "outsiders within," thinkers socially positioned to see the world "from margin to center," and from the "borderlands."[5] Global feminist approaches seek to explain the role of gendered institutions, practices, and cultures in the global political economy. They want to show how such institutions depend upon the maintenance of masculine cultures in militaries, governments, and multinational corporations, as well as among their allies in national and local community, such that the continued appropriation of women's labors and resources in different ways in every part of social relations can be organized and legitimated. They have challenged the adequacy of liberal and Marxian analyses of world systems, showing the commitments to male supremacy as well as eurocentrism that have also shaped such accounts.

The rich and complex analyses of these distinctive postcolonial feminisms contribute to and challenge earlier feminist analyses in at least the following three ways. They focus on the resources and limitations to be found in the different cultures within which women struggle against male supremacy, cultural imperialism, and economic immiseration. They focus on analyzing the mutually causal effects of changing gender relations and changes in global political economies. And they focus on how the eurocentrism of northern assumptions exacerbates the effects of each culture's existing androcentrism, leading to further deterioration of women's resources; eurocentrism is also androcentric. These postcolonial analyses beckon feminist science and technology studies with the powerful resources they obviously have provided to research and scholarship in literature, history, cultural studies, economics, and other fields. Many feminist science scholars teach these writings in our feminist theory or feminist epistemology courses; they are present in the introductory women's studies and feminist philosophy texts that we assign to freshmen. And our students at most U.S. universities increasingly are themselves multicultured and are increasingly aware of the distinctive cultures of their own family inheritances. Every day they become more aware through their other courses, the media, and other public discourses of the intellectual and cultural traditions of non-European cultures and of the ways eurocentrism has devalued and suppressed them.[6]

Contemporary northern feminist science theories have already begun to think about gender issues in scientific and technological change from the perspective of such concerns, though it has been more difficult to see how to approach such a project in abstract theories such as epistemologies and philosophies of science—as usual. A major obstacle to the feminist approaches here is the eurocentrism of conventional philosophies, histories, and social studies of science and technology within and against which the northern feminist analyses have been constructed.

Feminist standpoint approaches (cited above) have called for identifying and using the unique resources of women's particular social locations in order to identify and provide maximally objective understandings of sexist and androcentric presuppositions shaping dominant institutions, their conceptual frameworks, cultures, and practices. The postcolonial studies described in earlier chapters can expand the conceptual framework within which feminist science and technology analysts have begun to think about how to use the resources postcolonial feminisms make available. Thus, feminist postcolonial science and technology studies can play an important role in creating further dialogue among post-Kuhnian northern science and technology studies, the related northern feminist science and technology studies, and postcolonial science and technology studies.

The next section identifies some fruitful directions postcolonial feminist science studies have already taken. Important work here already has been developed by feminist scholars from many parts of the world, including Europe and North America, that situates feminist concerns about sciences and technologies on new historical and geographical maps that have been developing since World War II in postcolonial studies. Section 3 then turns to identify some contrasts between the conceptual and political frameworks of northern feminist science and technology studies and those that shape postcolonial feminist science and technology concerns. Subsequent chapters examine the conditions that generate these contrasting feminist concerns and some epistemological implications of this situation.

2. *Postcolonial feminist science and technology studies: Recent directions.* Studies that begin working up their science and technology issues from the lives of women in diverse cultures in global political relations have already moved forward in innovative and influential ways. Two important ones analyze how sexism has intersected with the racist and imperialist scientific agendas of northern "orientalism" and rethink the theories, policies, and practices of post–World War II development.

Gendered scientific orientalism. In the early nineteenth century, a biosocial science of racial and sexual difference developed that was based on a widespread analogy between race and sex/gender. Historian Nancy Leys Stepan explains how widespread was this practice of using sex/gender difference to explain racial difference, and vice versa.

> [W]hen Carl Vogt, one of the leading German students of race in the middle of the nineteenth century, claimed that the female skull approached in many respects that of the infant, and in still further respects that of lower races, whereas the mature male of many lower races resembled in his "pendulous" belly a Caucasian woman who had had many children, and in his calves and flat thighs the ape, he was merely stating what had become almost a cliche of the science of human difference.[7]

Stepan shows how it is not that sexism and racism have been two parallel discourses; it is not that sexism and racism just happen both to be characteristics of dominant institutions and their conceptual frameworks. Rather, each has been used to construct the other—at least since the early nineteenth century when gender difference began to join race difference as an object of scientific scrutiny. The phenomenon Stepan describes is a form of orientalism, memorably delineated by Edward Said.

> Orientalism depends for its strategy on this flexible *positional* superiority, which puts the Westerner in a whole series of possible relationships with the Orient without ever losing him the relative upper hand. And why should it have been otherwise, especially during the period of extraordi-

nary European ascendancy from the late Renaissance to the present? . . . [T]here emerged a complex Orient suitable for study in the academy, for display in the museum, for reconstruction in the colonial office, for theoretical illustration in anthropological, biological, linguistic, racial, and historical theses about mankind and the universe, for instances of economic and sociological theories of development, revolution, cultural personality, national or religious character.[8]

Said's account reveals how the discursive position that the Orient occupied in occidental institutions, their conceptual frameworks, cultures, and practices has been for centuries a gendered one even though the scrutiny of sex/gender difference emerged as a full-fledged scientific field only more recently. The Orient was the exotic but inferior feminine Other for the Occident's manly activities of study, authoritative museum display, colonial administration, natural and social science theorizing, and more.

Similar themes have been explored by biologist Anne Fausto-Sterling and historian of biology Londa Schiebinger.[9] Both discuss the shameful case of the "Hottentot Venus," a woman from southern Africa who was brought to Europe for scientific investigation and displayed naked to Europeans. Upon her early death she was dissected by the famous French naturalist Georges Cuvier, who then bequeathed her celebrated genitalia to the scientists at the Museum of Natural History in Paris. Schiebinger has also shown the inextricable weaving together of race and gender anxieties in the constitution of modern biology, especially during the heyday of eighteenth- and nineteenth-century European imperialism and colonialism. Gender traits persistently were invoked to explain purported racial superiority, and race to explain purported gender superiority. As more and more Europeans travelled through colonial Africa, apes made more and more frequent appearances in these accounts as the proof of whatever was being claimed.

Donna Haraway has produced a series of studies of the constitution of scientific fields such as primatology throughout the twentieth century through reliance on race, sex/gender, and class analogies for sexual reproduction, families, and dominance relations. Her work has also focussed on how primates have been conceptualized in these sciences as a kind of simple form of human society, more suitable than modern western ones for the scientific study of sexual reproduction and kinship relations.[10] In her classic early study, she looked at the union of the political and the physiological in animal sociology—the science of animal groups—and in particular at the work of Clarence Ray Carpenter. In the 1930s, Carpenter cofounded the first major research station for free-ranging monkeys as part of the school of tropical medicine, located on Cayo Santiago off the coast of Puerto Rico and affiliated with Columbia University. Carpenter's project was part of the field of dominance studies in the increasingly influential science of human engineering. Carpenter, a former student of Robert M. Yerkes at the Laboratories of Comparative Psychobiology at Yale University, followed his mentor's lead in believing that the social-sexual life of primates was thoroughly intertwined with their intelligence. How male dominance worked, how it could be created and, importantly, eliminated was the focus of these studies of animal societies that "were seen to have in simpler form all the characteristics of human societies and cultures" (230). Carpenter's studies showed how to link social functionalism to physiological functionalism in ways that would provide the framework for the ideologies of social control and techniques developed for, at first, medical, educational, and industrial management, and by World War II military and colonial administration. While primatology might appear to be only about animal communities, sex/gender, and dominance, the widespread location of such studies within institutes of tropical medicine or in field work in developing countries, the historic accounts of Africans as next to primates in the great chain of being, and the use of functionalist anthropological accounts to structure colonial administration all establish at least the early primate studies as part of orientalist colonial projects, among their other uses.[11]

In these studies of the sex/gender dimensions of orientalism, postcolonial feminism unmasks the political projects of sciences co-constituted by androcentric and eurocentric projects, their institutions, cultures, and practices. This kind of project is also part of the broad canvas of feminist examinations of so-called development.

Feminist critiques of "development." Ester Boserup's influential 1970 analysis of *Women's Role in Economic Development* marks "the date of the development 'discovery' of Third World women's poverty as well as their role in economic productivity," as a critic of the early "women in development" movement puts the point.[12] "Women in development" names the first feminist criticisms of post–World War II development policy, which argued that women, too, should get the development "benefits" that their brothers, fathers, and sons were receiving. However, these later postcolonial feminists argue, this strategy failed to identify the fundamentally bad effects development policies already were having on those who were already economically and politically most vulnerable. It was not only the mainstream development policies and practices that continued colonialism by other means in the name of development; initial feminist interventions in fact simply replicated the orientalism of the colonial discourses and their mainstream development heirs. Development policies were destroying the environments and cultures of the "unaligned nations" that had become the "Third World" during the Cold War period. Women were not at all left out of the narrow, economistic, development planning of the early postwar years, as a series of such postcolonial accounts countered. The appropriation of women's labor, land rights, and natural resources has been a crucial generator of profit for elites in the North and their allies in the South over the entire last half century of development policy.[13]

Modern science and technology are implicated in such critiques of economistic development and the feminist development analyses since development had been conceptualized narrowly as increased economic productivity, bereft of concern for the preservation and improvement of natural resources, local community relations, non-western cultures, or women's conditions. Increased economic productivity was to be achieved through the transfer of those northern scientific and technological information and techniques that are assumed to be responsible for the growth of European and North American economic productivity in the nineteenth and early twentieth centuries. Thus, the postcolonial critiques focus on the costs to the South of the systematic ignorance about nature's order and how to observe and explain it that are characteristic of development projects. The powerful feminist analyses in the more recent "women and alternatives to development" school show how ignoring women's relationship to nature on the one hand, and to the welfare of communities, on the other hand, insures the failure of development projects that is an increasing source of alarm in the North as well as the South.

These focuses of feminist postcolonialism on sciences and technologies in development will be pursued further in a later chapter. They are noted here as part of this preamble to looking at the partly contrasting conceptual frameworks that have shaped feminist science and technology studies within northern and southern feminisms. The gender and science discussions in the North have been developed within the liberal and socialist feminist political philosophies that have framed European and North American public agenda feminist demands on the state and public institutions. The postcolonial science and technology discussions mentioned above have been developed within political philosophies designed for different geographical and historical contexts. How do such differences shape thinking about gender, science, and technology?

3. *Contrasting frameworks. "Add women of color and stir?"* The concerns of postcolonial feminisms cannot simply be added on to feminist analyses that started off thinking about women's lives and the dominant conceptual frameworks in other times, places, and/or locations in global political and economic relations. It is not that nothing of value can be learned by trying to do so. Rather, conceptual frameworks constructed to accommodate the understandings of gender and science that have seemed most reasonable from the standpoint of the lives of women in

northern science and technology studies and those other women with whom they interact cannot help but leave unrecognized important issues for women thinking out of postcolonial realities and discussions.

The frameworks of northern feminist science and technology studies, no matter how much their advocates actively struggle against racism and eurocentrism, will systematically block even raising some of the issues most central to postcolonial feminisms. This phenomenon is familiar from feminist studies more generally. Most of the northern feminist theories have criticized the attempt simply to "add women and stir" to theories about, for example, rational man and the kinds of political, ethical, and epistemological reasoning he will employ, or the objective observer and the kinds of methods he will use to detect distorting social elements in research processes. It was not just that conventional theories of rationality and objectivity inadvertently neglected to talk about women, too, as rational and objective humans. Worse, these concepts were often formulated precisely in opposition to the feminine. Prevailing northern conceptions of rationality and objectivity, in spite of their great diversity throughout history (more than two millennia of history, in the case of rationality), evidently have virtually always been part of the construction of distinctive forms of masculinity. That is, what counts as rationality or objectivity is only what can be given a masculine meaning, and then the masculine is uniquely identified with the distinctively or ideally human. Men's preferred styles of reasoning or standards for maximizing objectivity thus have come to count as rationality and objectivity per se, leaving women's typical and valuable styles and standards marked as infantile, not fully human, or not ideal.[14] Thus, one can expect "add women of color (or some feminist form of postcolonialism) and stir" to face the same kinds of problems encountered by attempts to add women and gender to conceptual frameworks designed, intentionally or not, to exclude them. Of course, postcolonial feminist frameworks will also challenge those approaches in postcolonial science and technology studies that do not see women's and feminist issues as central to understanding how colonialism, postcolonialism, and sciences and technologies work.

What are some of the most obvious contrasts and even sometimes conflicts between northern and southern feminist science and technology conceptual frameworks? I point here to tendencies more than to sharp contrasts. Many individual scientists and science studies researchers trained in northern institutions and their conceptual frameworks have been thinking in fresh ways about resources that postcolonial feminist scholars generate. There are increasing opportunities for them to meet and dialogue with feminists working in formerly colonized cultures. And many of the latter, whether originally by birth or training professionally "from" northern or southern institutions with their distinctive conceptual frameworks, have been in dialogue for decades with northern feminist science and technology studies. So the following suggestions are intended not to repolarize positions that in fact are learning from each other, but rather to show the importance of the multiple locations in economic, social, and political relations around the globe from which feminist science and technologies studies are emerging.

Starting from southern lives. First of all, northern feminist science studies start off thinking about their topics from the standpoint of the lives of women of European descent in the North, and the institutions, conceptual schemes, and practices that they experience as most exploitative and dominating. This is a valuable site from which to examine modern sciences and technologies as they do, since it is in the North that the prevailing idealized representation of modern sciences, their accomplishments, and processes was created, and where it still most powerfully shapes people's thinking in scientific and scholarly disciplines and in public life. Moreover, it is this representation of science and of gender relations that is becoming increasingly powerful around the world, so it is crucial that feminists "at the center" criticize it. However, there are other aspects of modern sciences and technologies, other scientific and technological traditions, and gender issues in both that cannot so clearly be perceived from such northern locations.

The postcolonial feminist analyses start off from feminist discourses about the realities of the lives of women of other cultures in the North (African American, Chicana, Asian American, Native American, etc.) and South, and of women's positions in the global political economy, for example, in the projects of multinational corporations, Third World development, and as victims of militarism. They bring different resources to projects of reexamining scientific and technological change from the standpoints of these lives. Of course there is no one single "correct" way to understand such starting points for research and policy. The realities of the lives of women of other cultures in the North and South, and of women's positions in the global political economy have been theorized in diverse ways that are opposed to sexist and imperialist conventional views. The debates between advocates of these different postcolonial standpoints are useful. The next chapter explores resources women can bring to knowledge-seeking projects from their different locations in such local and global social relations. The ability of northern and postcolonial feminisms to make distinctive contributions to the storehouse of human understanding about nature and social relations, including sciences and science and technology studies, depends upon the ability of each to make use of the local resources at their disposal.

The point here is that postcolonialisms raise distinctive issues that have not emerged, or have emerged only in different forms, in northern feminist science and technology studies.

A role for postcolonial histories and geographies. Second, postcolonial feminist science and technology studies begin from a different history and geography of European and other sciences than do the northern feminist studies. Much of the latter share with the post-Kuhnian northern science and technology studies a eurocentric kind of segregated and "nativist" perspective on the history of science. It is segregated in that modern science is conceptualized as having little or nothing to do with the history of European expansion in the past, with contemporary Third World development policies and practices, or with the stagnation and, often, destruction of science and technology traditions of non-European cultures. It is "nativist" in that the perspective of the "natives" of European scientific and technological cultures on their culture, its history, and present practices is taken as authoritative. The "natives" here are not only scientists-in-their-laboratories, but also those who have selected and funded its projects, and those who have received most of its social and material benefits and relatively few of its costs. Earlier chapters showed the importance of the post–World War II emergence of postcolonial histories and their science and technology studies for understanding the history of modern European sciences as well as of other cultures' knowledge traditions. Postcolonial feminist science and technology studies are located on a different historical and geographical map than the northern feminist studies that restrict their attention to the European/American scientific and technological tradition, as that has been understood within the framework of eurocentric histories of modern and other knowledge traditions.

Institutional priorities. Third, a main part of postcolonial feminist studies emerges from different institutional contexts than the northern studies, and these institutional origins also shape their priorities. In the United States, at least, the northern feminist accounts have been produced primarily in the contexts of university research and teaching, struggles in natural and social science disciplines, research sites, funding organizations, and dissemination outlets such as journals and conferences, in the philosophies of natural and social sciences, and in policy struggles about women's health, reproductive technologies, workplace technologies, and northern ecological/environmental issues.[15] In the United States, science studies, on the one hand, and technology studies and Third World development studies, on the other hand, are institutionally highly segregated. The former are conducted in arts and sciences schools in history, sociology, anthropology, philosophy, literature, and biology departments. The latter are produced in engineering, agriculture, medical, and public health schools. Arts and sciences scholars do not usually think they have anything to learn from applied science and technology researchers—an issue discussed further below.

In contrast, central components of the postcolonial feminist science and technology discussions have been produced primarily in the context of struggles against and within international agencies, local patriarchies, and grassroots organizing in the South. These analysts often are trained or located in the applied science and technology schools. The local is the object of their expertise. With different priorities, northern and postcolonial feminist studies will tend to produce different patterns of science studies knowledge (and ignorance).

At issue here are kinds of institutional, social, and philosophic eurocentrism. In science and technology studies, as in scientific and technological fields themselves, prevailing patterns of research priorities and standards of adequacy are only marginally determined by the actions of individuals. Instead, institutional standards for what counts as intellectually, politically, and practically fruitful research are set implicitly and explicitly by institutions, their cultures, and practices. These institutional standards are often aligned with larger social standards and sometimes even with broad-scale "civilizational" or philosophical standards that span many cultures and/or historical eras. For example, when graduate schools, disciplinary journals, and conference organizers assume that "science" refers only to what physicists and chemists are willing to think of as model sciences—or even as sciences at all—then many science and technology issues central to the lives of women, especially in cultures that have relied on their own non-western scientific and technology traditions, will not even be seen as relevant issues for feminist science and technology studies. These institutional practices have found support in the larger society in the United States and through long stretches of the history of science in the United States.

Different concepts centered. Finally, the concepts centered in northern and postcolonial feminist science and technology studies analyses differ in significant respects. Let us look at just a couple of these contrasting concepts.

i. What is gender? Multicultural and global feminisms need richer and more accurate accounts of gender than those that have tended to be found useful in northern science and technology studies. Of course gender differences are not a natural product of sex differences, as is still widely believed in many cultures. Gender is socially organized and evidently plays an important role in the cultural construction of what counts as real sex differences.

Multicultural and global feminist accounts have been one important source of the shift from thinking of gender as fundamentally a thing or property attached to individuals to thinking of it as a relationship that is fundamentally an attribute of social structures and symbolic systems.

Though the content and meanings of masculinity and femininity vary from culture to culture, the duality itself evidently is a continuing resource for the way different cultures distribute scarce resources through the organization of their institutions and practices and give meaning to the apparent disorder of nature and social relations through symbolic systems. Thus, gender is a property not only of individual men and women, but also of such social structures as divisions of labor and principles, for distributing social benefits and costs, and of meanings. For example, analysts speak of how Third World development policy and practices are (masculine) gendered and of the androcentric meanings of the rational economic choice theory that shaped the definition of what would constitute development after World War II, as a later chapter will explore.[16] Moreover, gender is always hierarchically organized—not merely complementary. Masculinity is always valued more highly and rewarded more richly; it is evidently always associated with the distinctively human and the most prized accomplishments of public life, as indicated earlier.

Furthermore, multicultural and global feminisms have been crucial to developing the argument that gender is always interlocked with class, "race," ethnicity, colonialism, sexuality, and whatever other hierarchical social relations organize a culture's discursive frameworks, including its material, institutions, and practices. Thus, these "macro" forces are mutually constructing and maintaining. They form a social matrix in which each of us has a determinate location (individual, structural, and symbolic) at the juncture of diverse hierarchical social relations.[17] Finally, gender relations are dynamic, not static. They are historically changing ways of obtaining and distributing scarce social resources; during any kind of social change they become sites of political contestation.

Many accounts in northern feminist science and technology studies have tended to lose sight of some of these important dimensions of gender relations. But without this kind of complex understanding, it is easy to miss seeing the significance *for women's lives in the North* as well as in the South, of the issues central to postcolonial feminist science and technology studies. Women's lives in the North are as fully part of multicultural and global social relations as are the lives of women and men of non-European cultures living at other locations in global economic, political, and social relations. Northern women cannot understand scientific and technological issues in our own lives to the extent that we fail to grasp the insights of postcolonial feminist science and technology studies.

ii. What do sciences study? Nature versus the environment. Northern and much of postcolonial feminist studies seem to rely on different conceptions of the object of scientific study in their analyses. "Nature" is the focus of the northern discussions (except for environmental ethics, which is not usually counted as part of science and technology studies), but it is "environments" that are centered in the postcolonial ones. "Nature," with all its meanings and their histories, arrives in northern feminist accounts from deployment in the distinctively European and U.S. discourses of Mother Earth, wild and unruly nature that man must tame lest he lose control of his fate, nature speaking the language of mathematics that must be learned by those who would discover and explain her regularities, chaste and resistant nature whose veils must be stripped so that science can discover her secrets, and the nineteenth-century U.S. invention of nature as "the wilds" to be revered in an imagined pristine state unsullied by human presences . . . and more. Feminists criticize these discourses, of course, but it is the nature shorn of such meanings that is the object of the philosophies of science that they develop and assume.[18]

In contrast, "environment" arrives from a different set of deployments in discourses about the surroundings with which humans regularly interact in daily subsistence struggles, that can be appropriated for multinational corporations' purposes, that can be destroyed by human arrogance, greed, and carelessness, that must be sustained if human and other forms of life are to survive. Where "nature" appears in southern feminist discussions, it frequently has a spiritual and moral dimension; it is not the "dead matter in motion" that is the object of modern scientific and technological observation and manipulation. There appears to be no culture-neutral object of scientific study since how such objects are conceptualized is a matter in part of the legacies of conversations local scientific groups call on in delineating what will and will not count as part of their objects of study.

iii. "High" versus "low" sciences. Moreover, a series of contrasts hinted at in the paragraphs above cluster in northern and southern feminist science studies. On the one side stand "pure nature" and the "pure science" that can tell the one true story of pure nature's order. On the other side stand various "impure" objects of knowledge and the disciplines or methods of knowledge seeking that try to understand them. Pure physics, chemistry, and formal biology are only rarely, if at all, the object of postcolonial feminist science studies examinations. In the postcolonial accounts, technologies often are of more interest than any sciences to which they may be linked. Sciences that study objects of mixed natural and social knowledge—such as health sciences, agricultural, and environmental sciences are far more the object of southern feminist interests. Such sciences must themselves combine theoretical frameworks and skills found both in the natural and social sciences.[19] Of course, northern feminist (and mainstream) studies point out that physics, chemistry, and formal biology also study objects produced by social interests, discursive resources, and so on, as well as nature's order. Moreover, the "mixed sciences" are also the object of northern feminist science and technology studies. But they are very much the center of the postcolonial studies and far less so of northern studies.

These differing priorities are to be expected when postcolonial feminist science and technology projects are generated in different institutional contexts. They also represent the distinctive positions in multicultural and postcolonial global social relations from which they have emerged.

4. Conclusion. Yes: feminist science and technology studies should be culturally diverse. Recognition of the value of different approaches can open spaces for further discussion that can enrich the thinking of all parties to such debates. Feminist thinking needs the creative postcolonial feminist projects that are helping to set past, present, and future scientific and technological projects more firmly in their historical era, while also mapping policy initiatives that can benefit women's as well as men's lives. Further, it also needs concerns and innovative analyses of northern feminist science and technology projects, which have trained the powerful lens of northern feminist theory onto the politically most powerful science and technology institutions, their cultures, and practices, that the world has ever seen. The next chapter explores how it is not only women's distinctive political locations that enable this kind of fruitful diversity in feminist standpoints, but also their different locations in nature and in cultural formations.

Such valued diversity also points toward the need for radically different epistemologies and philosophies of science than the prevailing mainstream ones. Such philosophies will have to be able to see such multiplicity not as a perhaps necessary but only temporary stage of feminist thought, but as an example of thought that at its best creatively responds to the natural and social transformations of its era. Such philosophical issues are pursued in later chapters.

6

Are There Gendered Standpoints on Nature?

1. "Gender cultures" and the production of knowledge. Post-Kuhnian northern science and technology studies and postcolonial southern science and technology studies both have shown how modern sciences are just one collection of local knowledge systems among many others that have existed in the past, are present in other cultures today, and will continue to be produced in the future, as earlier chapters reported. It is strange to think of modern sciences as local knowledge systems when it is just such local systems with which prevailing idealizations of scientific processes as universal ones conventionally have been contrasted. Nevertheless, focussing on how local resources have constituted scientific projects illuminates aspects of scientific and technological change that otherwise are obscured or inexplicable.

Of course in many respects modern sciences obviously are much more powerful cognitively and politically than older European knowledge systems and than the knowledge systems of other cultures. However, other such systems have been able to learn a lot about the natural world before modern sciences, and even that modern science has not yet learned; all the achievements are not those of modern sciences first or alone. Earlier chapters showed how modern sciences bear distinctive marks of the local resources available to them from their base in Europe and the North. These resources have functioned as toolboxes for the growth of systematic knowledge in some respects and as prison houses that have produced systematic ignorance in other respects.

This chapter adds the voice of a third important school of post–World War II science and technology studies to such an account. Cultures' gender formations—their gendered social structures and symbolic systems—are also local resources that have had both productive and regressive effects on the growth of knowledge, as several decades of feminist science and technology studies have shown. Of course gender formations differ from culture to culture, just as do class, race/ethnicity, religious, and other institutionalized social relations. This chapter explores a conceptual framework for analyzing the resources that women's distinctive standpoints on nature can provide for science and technology policy and for the social studies of science and technology.[1] The next chapter examines some of the resources that postcolonial feminist science and technology studies, such as the women, environment, and alternatives to development movement, have provided for enlarging and deepening productive dialogues between all three approaches to rethinking scientific and technological change.

Chapter 4 examined how it is more accurate and useful to understand different cultures as having distinctive relationships to nature because of the local resources provided them by their geographical locations, and their interests, discursive resources, and culturally distinctive ways of organizing the production of knowledge. In that discussion cultures were treated as internally gender homogenous—as if they were composed of humans of no particular gender. But, of course, they are not. Women and men in the same culture have different "geographical" locations in heterogeneous nature, and different interests, discursive resources, and ways of organizing the production of knowledge from their brothers. Here the focus is on gender differences, on the reasons why it is more accurate and useful to understand women and men in any culture as having a different relationship to the world around them.

Many readers will find it strange and objectionable to consider the possibility that there are such things at all as gendered standpoints on nature—women's and men's distinctive relationships to the natural order. After all, don't women do the same science as men in biology, physics, and chemistry labs? Hasn't feminism taken a firm stand against the myth that women can't do "real science," or can't do it as rigorously and creatively as can men? However, the argument here will be that, nevertheless, analyses of how it is that women's lives can generate resources different from those arising from the men's lives upon which modern scientific and technological thought has been based deserve more appreciation than many of us, feminist and prefeminist alike, have given them. The issue is not only one about understanding women's past achievements, or only women's knowledge in premodern societies; it is also about the resources that starting off research from women's lives can provide for increasing human knowledge of nature's regularities and their underlying causal tendencies anywhere and everywhere that gender relations occur. It is also, in part, an issue about women making science policy, whether or not they are practicing scientists.

Before turning to the chapter's topic, a brief word is in order on the relationship between the analysis of this chapter and feminist standpoint theory that examined the scientific and epistemic resources made available by starting off research from the lives of people who have been disadvantaged by the dominant conceptual framework—in this case, women. In an earlier chapter I pointed out that a "Europology" had two analytically separate but historically entwined components. Such an account would show, on the one hand, that European thought is distinctive because it uses local resources, as does the thought of all other cultures. Its interests are the interests of Europeans; its metaphors, models, narratives, and discursive resources are those of European history—not of Asian, African, or some other history. Its methods and styles of thinking are similarly distinctive. On the other hand, its distinctiveness is also to be found in those aspects of its content and conceptual framework that were and are enabled by its dominating and exploitative location in global power relations. Of course these two aspects of European thought can only analytically be distinguished since in practice the content, form, and uses of Europe's local resources are understandable only within the global power relations in which Europe engaged and which stimulated European fears, hopes, and fantasies. The Europology provided by postcolonial science studies shows a view from "outside" European local resources and also from "below" in global postcolonial relations. It is shaped, so to speak, by differences both between simply distinct cultures and between ones in hierarchical relations to each other.

Most analyses that have focussed on "difference" have tended to focus on either "mere difference" or hierarchical differences, but not on both. On the one hand, focuses on power relations tend toward binarism, an overemphasis on discrete categories of "the powerful" and "the powerless," and the homogenization of differences within those categories. On the other hand, studies of different cultures often tend to lose sight of the global political economy and the unequal relations it creates between cultures, as well as of pervasive power relations such as gender relations that create similarities and alliances between, on the one hand, those who can exercise economic and political power and, on the other hand, those who are the object of others' power exercises. The standpoint of "others" is a good place to start thinking simultaneously about cultural difference and power relations. It can generate insights about how both kinds of difference contribute to enlarging and limiting knowledge of the natural world.

Feminist standpoint theory mapped the different scientific and epistemic resources that hierarchical gender relations create for starting off research or thought from women's lives as the latter have been understood in various feminist discourses. It showed how positions of political disadvantage can also be turned into sites of analytic advantage. This chapter steps back for the moment from the gendered power relations that were the central concern of earlier standpoint theory to examine the distinctive resources that can be accessed by starting thought simply from whatever distinctions have been socially or biologically created between women's and men's lives. That is, women's lives would be valuable places from which to start off thought even if gender were not fundamentally a political relation, like, in this respect, class and race. Of course gender is indeed fundamentally a political relation (as we turn shortly to examine further), so in everyday life mere gender difference does not exist; power relations are always in play here, subtly or not. Nevertheless, this analytic exercise seems interesting and valuable to pursue. It can be structured parallel to the "different cultures" arguments of the last chapter, in effect treating gender differences as if they were differences between cultures—as if women and men in the same historical and geographically located culture nevertheless lived in different cultures. Another way to put the point here is that this chapter tries to fill in a structure for understanding how the "mere differences" in men's and women's activities and the meanings of manliness and womanliness make women's knowledge a distinctive human resource quite apart from advantages provided by the recognition of its suppression.

What can we learn when we think of women and men as living in different cultures?

2. *Gendered cultures?* Until fairly recently, most people thought of gender differences as properties of individuals that were entirely an outcome of biological sex differences.[2] Many people still hold this older belief, and in many languages there is no word for gender. In the last several decades, more accurate and useful conceptions of gender relations have been developed, as the last chapter recounted. To recap briefly, such accounts show how gender is fundamentally a way of organizing social structures and meaning systems; individual gendered identities are a consequence rather than a cause of gendered social relations and meanings.[3] Gendered social structures are produced by assigning some work and activities to women and others to men. Gendered meaning systems are produced by using putatively observed contrasts between women's and men's characteristics to give meanings to natural and social phenomena that have little or nothing to do with biological sex differences. Thus, gender is neither an outcome of sex differences nor a matter only of how each of us is identified by ourselves and others.

Moreover, as we saw, gender is fundamentally a relationship, or series of them. In order accurately to describe and explain women's lives and the meanings of womanliness we must look at the relations between women's and men's lives and between womanliness and manliness. In different cultures, gendered social structures and meanings take on distinctive characteristics: what is manly in one culture, such as engaging in public sphere economic relations, is womanly in another culture, such as west Africa where women are the market traders. Gender relations are to some extent or other always hierarchically organized since men tend to control more social resources and manliness is more closely identified with the admirably "human." Moreover, gender relations are always co-constituted by/with other hierarchically organized social relations such as class, ethnicity, race, and so on. An African American man's life is shaped by the gender relations peculiar to his position in racial as well as gender relations, and the class, ethnicity, sexuality, and other social relations that structure his particular society. Similarly, white women's opportunities and responsibilities are shaped by the way their gender location, meaning, and identity is co-constituted by class, racial, imperial, sexual, and other such organizing forces in a culture. Last, gender relations are never static or fixed; they are dynamic, historically changing ways of organizing access to scarce social resources. Thus, every other kind of social change offers opportunities for creating new forms of gender relations. To put the point another way, every kind of social change is always also a site for struggles over gender relations. Once we have in mind this richer and more accurate account of gender relations, it becomes easier to identify the ways in which women and men live in something like "gender cultures," and how gender struggles take place at moments of scientific and technological change.

Yet it may still seem confusing to speak of women and men living in different gendered cultures. Of course, in an important everyday sense, women and men live in the same cultures. One familiar way to specify cultures is by ethnicity, nationality, or some other category that contains both women and men, as in a "Native American culture" or a "French culture." However, sociologists, anthropologists, and historians also study cultures of street corners, locker rooms, schools, laboratories, corporations, and militaries. Some of these cultures are in fact completely or highly gender-segregated and carry highly gendered meanings even when individuals of the other gender are visible and active in them. Elementary school classrooms are thought of as women's cultures in recent U.S. history, even though plenty of men teach in them. Militaries are usefully perceived as masculine environments even when plenty of women are part of them. It is in this sense of "culture" that it becomes illuminating to ask how women and men live not fully in the same culture, but in partially different ones.

Ironically, readers of this book are probably in the group most unlikely to agree with the idea that they live in a culture different from their kin, friends, and colleagues of the other gender. Middle-class women and men may well be most resistant to the idea that they live in different cultures—more so than men and women in lower or upper classes. In western middle classes today, men and women tend to share many more androgynous life resources than they did in the recent past. After all, liberal feminism has always been designed primarily to give middle-class women equal access to the benefits available to their brothers. It is in the middle classes that in the late twentieth century are found the greatest similarities in women's and men's educations, career opportunities, incomes, and responsibilities for family life. And it is clearly here that women's aspirations toward "equality-as-sameness," toward eliminating gender-differentiated cultures, have been strongest.

It must be noted that such tendencies toward androgynous life opportunities may have begun to spread to the working class out of necessity, not choice. The effects of the worsening economy on working-class families in the United States has led to these men's greater participation in childcare and domestic work as their wives must bear more of the burden of providing family income. Moreover, the jobs available to working-class men increasingly have been de-skilled and turned into "McJobs"—to the kinds of temporary and part-time work formerly more characteristic for working-class women.[4] The point here is that wherever androgynous life-resources are found, there will be at least less structural gender since one major way of constituting gender difference is through a structural division of labor or activity.

In none of these cases would it be correct to say that life patterns are by any means identical for women and men, or that women's struggles in these classes are over. The issue here is that women and men scientists, university faculty, and students tend to be in the groups that find it less plausible to think in terms of different gender cultures. This class not only in fact has more androgynous life resources, but also has been trying to make available to women, too, the resources of education, financial opportunity, jobs, and the like that are available to middle-class men. So we, who are in this group, should consider the probability that our resistance to the idea of gendered cultures may be a product of our politics and our distinctive class situation.

3. The specter of women's sciences. Thus, for professional class women and men, the idea of women's distinctive standpoints on nature may call up absurd images of women's sciences. The reasons for such responses are not hard to detect and need to be acknowledged at the beginning of my argument here. In the first place, even raising such a topic risks expectable "misreadings" of it as (falsely) asserting that women's biology leads them to think differently—that there are genes or hormones or brain organization for sex-different sciences. However, even if there should turn out to be "biological" differences in the ways women and men think, other differences are my focus here. The claim that women are likely to have different resources for understanding nature is neither reducible to nor dependent upon claims that their brains are different.[5]

Second, the idea of "women's sciences" raises the specter of the horizontal discrimination that restricted the job opportunities of fully credentialled women scientists to careers in cosmetic chemistry and home economics, and in medicine to pediatrics.[6] In many countries today, women scientists tend to be clustered in a small number of specialities thought of us women's fields. The causes of this clustering are to be found not only in stereotypes of appropriate women's and men's activities and concerns, but also in the vagaries of local cultural and scientific history. For example, men often abandon a research field when the pay and status begin to be higher in others, and thus create women's sciences by default.[7] I do not mean to suggest that it is unfortunate for the rest of us that pediatrics, nutrition, public health, and other such socially valuable research fields attract and retain women scientists, but only that the idea of distinctive women's standpoints on nature can call to mind patterns of career discrimination against women.

However, in the third place, feminists have been troubled by other issues such as the way claims for distinctive women's standpoints challenge the dogma that the only good or real sciences are culturally neutral ones. Feminism is supposed to be removing sexist and androcentric cultural biases in the sciences, not substituting other cultural presuppositions, values, and interests for them, the criticism goes. If good science is not culturally neutral, then a host of unattractive consequences seem to follow. After all, women have claimed the right to equal access to scientific educations, jobs, publication, membership in scientific societies, and scientific awards on the grounds that they can do precisely the same, just as good, objective, value-neutral science as can their brothers. It is precisely women's transcendence of their womanly, "feminine" characteristics that have earned them their hard-won places in the sciences. Equality has meant sameness with men for these feminists, as it has had to mean for all of us dealing with cultures where it is not just a linguistic accident that "man" and "human" routinely can be used interchangeably. Moreover, to venture beyond the claim that women can do just as good science as men not only lends legitimacy to the long-standing, powerful devaluation of women's beliefs and practices; it also devalues the status of modern sciences by implying that they represent not distinguished human achievements, but only mere masculine ones. Such a claim seems, and largely is, absurd. Sciences "work" precisely because they enable prediction and control of nature regardless of the gender of the scientists making such predictions.

Fourth, to many feminist thinkers in the social studies of science, the question of women's standpoints appears inextricably mired in various contested assumptions. They think it essentializes women, centering some idealized "woman" or set of them, or some purportedly typical woman's activity, such as mothering, as representing all women and the womanly. Or it directs us to conceptualize scientific authority as grounded in something very different from universally valid knowledge claims, namely the beliefs about nature of each individual woman in the world—a reductio ad absurdum of claims for women's standpoints. These are misreadings of standpoint arguments, but they lead to resistance to the idea of distinctive women's standpoints on nature.[8]

Finally, there are those who think that such a question distracts from more important ones of whether there can be not women's or feminine sciences, but feminist ones. The question for feminists should be not only about what scientific fields or styles of doing and thinking have been permitted or encouraged in women living in male supremacist societies, but, more importantly, about what kinds of sciences feminists should want in order to end male supremacy and to increase women's access to the resources they need to live, and to live well. The question should be not about women's standpoints, but about feminist ones.

Readers will find some of these issues more compelling than others. Collectively, however, it is clear that the terrain marked out by my question is a confusing, difficult, and dangerous one. However, the argument here is that this is not a good enough reason for feminist or for post-Kuhnian or postcolonial science and technology studies to set it aside. The next section shows how the four kinds of differences between cultures identified in an earlier chapter also mark differences between the resources available to sciences and technologies that frame their projects from the perspective of men's or of women's lives.[9]

4. Women's standpoints on nature. Scandalous though gendered standpoints on nature may appear to some, the topic certainly is not new in feminist science and technology studies that are familiar in the North. Women's bodies are one object of scientific scrutiny about which it has seemed plausible to claim that there are such distinctive women's standpoints.[10] Another is the environment.[11] Indeed, some have read Evelyn Fox Keller's work on Barbara McClintock as making such claims about the far more abstract topics of genetics and about scientific processes.[12] But what is it about women's situations that could or does make possible these standpoints? Is it biology? Culture? Can men have or take these standpoints? Do all women have them? Do all women have the same standpoints? What about women scientists? In general, in what ways are women and men located in gender-distinctive cultures within their national, ethnic, or other historically located cultures that both enable and limit what can be known about the natural world from the perspective of their lives? Our answers to such questions will follow the pattern laid out in chapter 4.

Heterogeneous nature. First, women and men in any culture interact with partially different natural environments. In many ways they are exposed to different regularities of nature that offer them different possible resources and probable dangers and that can make some theories appear more or less plausible than they do to those who interact only with other environments. To start with, obviously females and males are "biologically" different,[13] though not in all those purported forms of such interest to biological determinists. Of course their reproductive systems differ. Menstruation, pregnancy, birthing, lactation, and menopause, to mention only the most obvious such processes, occur in women's bodies only. However, women differ from men biologically in other ways: in percent body fat (affecting, for example, their toleration for cold water in long-distance swimming), skeletal construction (affecting their abilities in sports, including their distinctive strengths and their distinctive susceptibility to injuries), susceptibility to the effects of drugs, and so on. Such biological differences expose them to different regularities of nature; they interact with some different aspects of nature than do men.

Moreover, socially assigned, gender-segregated activities—from tending children, the aged, and sick, to other unpaid domestic labor, local community maintenance, clerical work, factory work, service work (in social service agencies and health care, in paid domestic work, in the sex and tourist industries, etc.), gathering food, and maintaining subsistence agriculture, herding, and forestry—bring women into yet more distinctive interactions with natural environments. Women's biology and their culturally distinctive activities enable them to have some different interactions with natural environments than those their brothers have. Research that starts out from women's bodies and interactions with nature, too—not just men's—will arrive at more comprehensive and accurate descriptions and explanations of nature's regularities. So we could say that women and men are partly differently located in heterogeneous nature; or, to put the issue another way, heterogeneous nature is partly differently distributed in men's and and women's lives. Of course nothing forces women or men (or us) to interpret such situations in any particular ways. We have not yet arrived in this analysis at the ways that people are conscious of such differences, let alone at the production of knowledge, but only at an observation about distinctive sex- and gender-differentiated locations in nature "in themselves," so to speak.

Gendered interests. Three other kinds of gender-differing resources combine to enable and limit how people's locations in nature are turned into culturally distinctive objects of knowledge claims. First, women's and men's interests in their bodies and environments can differ. People tend to be interested in what they find around them. (Of course how they characterize and, thus, see "what they find around them" is discursively constituted as we shall see in a moment.) Clearly, women and men have at least partially different interests in and thus desires about the natural world to the extent that they are biologically different and that they are assigned activities that bring them into systematically different interactions with their environments. For example, women need and want to know how our bodies work, how drugs affect us, too, and how more effectively to interact with the environments on behalf of our own interests, including the well-being of the children, household members, kin, elderly, and sick, and the community networks that depend on women and for whom women usually are assigned responsibility.

Of course both men and women have children, other kin, sick and aged people around them, and live in communities; but they have at least partially gender-distinctive interests in and desires about these people's situations. These vary from culture to culture; each culture has norms for what are appropriate interests and desires for women and for men. These interests are most segregated where adult activities are maximally segregated. For example, there is geographic segregation in cases where men work in distant cities, in migrant agricultural labor, or in other countries as guest workers, only rarely returning to the areas where their wives and families live. Gender interests can even be highly geographically segregated where professional-managerial men spend long hours on the job downtown or travelling, and little time in domestic work or in the suburban communities where their wives and children live.

There are many other contexts for thinking about women's differing interests in nature. For example, recent reports show that the vast majority of fatalities in war are not military personnel, but civilian populations and, especially, women and children. Moreover, a much greater proportion of women's life work than men's is invested in birthing and raising the next generation, including the sons who will become soldiers. It is implausible to presume that women have the same interests and desires as their brothers to support militarism and the scientific and technological research agendas that service militaries. The point here is not that women's essential "natures" lead them to greater pacifism—that women are inherently more peaceful (they are not)—but rather that they bear more of the costs of warfare. The presumption that such sciences represent "human interests" is even more implausible when one considers how ideals of the warrior and the national hero are ideals of masculinity; women's compensation for the costs of war is far less than men's since they share only vicariously in the masculine glory that accrues from military victories. Successful military careers fre-quently generate "capital" that men can use to gain civil rights, public office, and other forms of political power. Women have fewer interests in research, basic and applied, that in fact services militarism and nationalism than do men, and more interests in research, basic and applied, that services sustainable domestic and local community activi-ties. Basic research is also an issue here, not just applied research, since the part of heterogeneous nature that interests a research community can vary according to culturally differing interests. For example, the choice of which bodily processes a basic research project focusses on can predictably be useful for responding to the interests of militaries, population control councils, genetic engineering corporations, or women's health advocates. Thus, women are repositories of historically developed knowledge about what is of interest to them. This knowledge is not static or "premodern"—an issue to which the next chapter returns. It always must be refined and developed in each generation as both nature and social relations, including gender relations, change over time. The environments that women interact with in their distinctive agricultural work, animal husbandry, trading, office and industrial work, child care, and responsibility for family health, for example, are all radically different now in many parts of the world than they were even two decades ago.

Of course interests and desires are not objectively given by any facts, natural or social. They are arrived at through culturally distinctive discourses—religious, national, ethnic, class, and are always historically specific. Thus, there is something artificial about separating out, as I have done, women's and men's locations in nature and different interests, as if these were in real life separable from the discursive traditions that generate interests and culturally distinctive ways of thinking, fearing, fantasizing, and caring about our locations in nature. Nevertheless, when one identifies interests as creating both resources and limits for the production of knowledge about the natural world, one can draw on the long and rich resources of sociology and political economy's discussions of historically differing interests. We know how to think about how people's different and often conflicting interests and desires result in different questions, problems, concerns. These kinds of analyses can also illuminate women's and men's different and often conflicting interests and desires with respect to environments.

Such considerations lead to recognition of two additional kinds of cultural difference that generate distinctively gendered patterns of knowledge and ignorance.

Gendered discursive traditions. Let us stay here with the same examples from this rich and complex category that were discussed in an earlier chapter—metaphors, models, and narratives. Women and men do not have the same relation to cultural metaphors, models, and narratives precisely because these frequently carry sexual and gender meanings. A central project of northern feminist science studies has been to critically evaluate just these kinds of scientific and technological discursive legacies.

Scholars have pointed out how our understandings of nature and of research processes are limited when the favored metaphors, models, and narratives are associated with ideals of masculinity. They point to the androcentrism of familiar conceptualizations of nature as a mother who is endlessly bountiful, or as a wild and unruly harridan who must be controlled lest man's fate be threatened, or as a shy maiden whose veils must be stripped away so that science can discover her "secrets." In medical discourses women's bodies have been modeled on a factory; when the machinery no longer functions, as in menopause, the factory has lost all of its value, according to this model. And men's fantasies of potency have structured the way they think of interactions with nature in many respects. Such models have in effect functioned as evidence for scientific claims.[14]

One can gain more accurate and comprehensive understandings of both nature and research if one is alert to the fantasy elements in such gendered discursive resources since they overvalue androcentric models of nature and modes of conducting research, turning gendered meanings into evidence for the plausibility of research claims. Women will tend to be more sensitive to cultural metaphors, models, and narratives that devalue womanly beliefs, traits, and behaviors; they more easily detect the distorting components of such models of nature. They sometimes seek out other ways of using cultural traditions. Or they feel less bound by them—they are less well-socialized into seeing nature or ideal research processes through such models of gender hierarchy that often are prevalent in dominant scientific discourses. However, while women tend to be more sensitive to such slurs on their character, beliefs, or behaviors, it is by no means exclusively women who can examine sciences' cultural traditions from a critical perspective on such androcentric gendered discourses. Men, too, can and do take the standpoint of women on just such issues. A number of male historians of science have also made important contributions here.[15]

Thus, women (and men who take the standpoint of women's lives) think within and about local discursive traditions, and they develop and revise them in gender distinctive ways. Such differences contribute to important opportunities for describing and explaining nature's regularities.

Gendered organization of scientific work. Lastly, the cognitive content of modern sciences and other local knowledge systems is shaped by culturally distinctive forms of social organization, and, especially, of work. Do women and men tend to organize work differently, and thus to organize differently the work of scientific and technological research? One well-known case is reported by Donna Haraway.[16] Japanese women primatologists routinely learn to identify individually hundreds of the primates they study, a feat that Japanese men find remarkable, and that is apparently unprecedented in Anglo-American primatology. This skill enables them to learn far more about the complex social relations in these species. Another example is reported by Sharon Traweek.[17] She recounts how Japanese women physicists, doubly devalued in the international scientific community by being both non-European/American and not men, have ingeniously developed valuable alternative resources for advancing their research projects. For example, they tend to retain far stronger links to their former classmates in U.S. and European graduate schools than do their Japanese male peers. Excluded by the male Japanese physics community, they maintain a place for themselves in the international physics community networks of their old school chums that their male colleagues abandon upon graduation. Consequently, their work can sometimes become more closely tied into the work of the prestigious international physics community.

There are also indications that women scientists tend to organize their laboratories, their choices of scientific projects, and their publishing strategies differently than do their male peers. According to a report in *Science,* anecdotal evidence suggests that women laboratory directors tend to reduce competitive relations between their assistants and students, while men tend to encourage them. This difference in attitude toward competition has been observed also in men's and women's choices of research topics. In one study, men seem to prefer the highly competitive "hot topics," while women tend to take a "niche approach" that enables them to "develop mastery and deal with a relatively limited number of colleagues who are interested in the same field." This approach appears to be at least partly responsible for the fact that while women publish less, their papers tend to be cited at a higher rate. One possible explanation of this situation is that "women tend to take more seriously the internal requirement to turn out very thorough articles rather than just turning out a lot of articles."[18] Scientific institutions are social organizations much like other formal organizations in their local cultures. Consequently, the literature on women in organizations can be especially revealing here.[19]

Insofar as social relations shape scientific activities, any and all of the feminist social science literature examining women's and men's typically different ways of participating in social relations can also provide information about how the genders participate differently in the social relations of doing science. Moreover, the world that medical and health, environmental, and other part-natural-part-social sciences examine also offers opportunities for women to bring to their studies the distinctive understandings of social relations that the social sciences have reported. Thus, in addition to the sociological literature on women in organizations, feminist sociology, anthropology, economics, psychology, and other social science literatures report gender-differing ways of organizing the production of knowledge.

This section has been showing that a wide array of literatures provide evidence for claiming that distinctive standpoints on nature are created by women's and men's "biologically" and socially assigned different interactions with nature, by gender-differing interests and desires that shape such interactions, in gender-differing relations to available discursive resources that shape interests and desires, and in gender-typical ways of organizing the production of knowledge. Northern feminist science studies argued that scientific knowledge is local knowledge, not transculturally human knowledge, in its forms of historic gendering. Thinking about the distinctive resources that different cultures bring to representing and interacting with environments provides a model for analyzing how "gender cultures" also systematically generate ranges of distinctive knowledge about natural and social worlds.

5. *Conclusion.* In conclusion, let us consider the implications of such analyses for the work of women scientists, for the nature of local knowledge systems, and for public policy.

First, with respect to the work of women scientists, let us begin by setting aside some of the invalid and false conclusions to which some have been tempted. One certainly cannot conclude that women's biological differences from men automatically lead them to think differently; that women should go into those sciences focussed on issues centered in women's lives; that only women can make significant contributions to fields characteristically thought of as womanly, or that all women in those fields automatically do; that sciences not focussed on these fields are only masculine sciences, not human ones;[20] that there are typical or essential, or fixed and unchanging, womanly interactions with nature that can define once and for all what women's standpoints are; that women automatically know what women's standpoints on nature are; or that this issue of women's standpoints should replace important questions about which feminist analyses can and should best frame particular scientific or epistemological projects. Many critics of feminist science studies have assumed that commitments to such claims characterize all feminist science scholars, while in fact they have at most been made by only a tiny minority of the feminists working in the diverse disciplines and social projects that make up this field.

Instead, the proposal here is that when women and men scientists can figure out how to ask questions about nature's regularities that originate in women's distinctive "locations in nature" also, and in the kinds of cultural resources that women's interests, relations to cultural traditions, and ways of organizing knowledge make available, then such sciences will produce information that better serves women, too; and the storehouse of human scientific knowledge can expand and gain in accuracy. Note that the point is that *women and men scientists* can do this—not just women. After all, women have had to learn how to ask interesting scientific questions from the perspective of natural and social locations that were not their own, for they were the social locations of the medical research establishment, or of militaries, or of industrial profiteers. Why can't men, like women, now learn how to think from the standpoints of women's lives? Of course, many already have. They have been able to step outside the dominant conceptual frameworks to locate them as simply the dominant ones, and to see other possibilities from the perspective of women's lives as these are delineated in various feminist discourses.

A second point is that from the perspective of this account, illuminating re-readings could be provided of the histories, philosophies, sociologies, and ethnographies of science provided by the last three decades of northern post-Kuhnian and the southern postcolonial science and technology studies that emerged alongside feminist studies after World War II. Both of these approaches to science studies have described in different ways the distinctively local character of modern sciences and technologies and, in the case of postcolonial studies, the scientific and technological traditions of other cultures. Culturally local European, Chinese, African, or pre-Columbian Andean patterns of knowledge and ignorance are also gendered patterns. Useful questions here could include: How have the reports of local histories, as well as the contemporary studies of them, been shaped by distinctive gendered standpoints? What can we learn by starting from women's lives—in those histories, and now—about scientific and technological changes in history?

Finally, the existence of gendered standpoints on nature has implications for science and technology policy and for public policy that depends on scientific and technological knowledge. Women's standpoints on nature provide resources for different and valuable ways to "see nature" and interact effectively with it. Sciences that do not use and help to develop such standpoints deprive themselves of the possibility of more accurate and comprehensive observations of nature and explanations of them. Humanitarian and environmental policy agendas that do not value and develop them insure their own failure. One example here will serve to preview some of these issues that will be pursued further in the next chapter. Observers have noted the striking fact that women have taken the lead in calling for and establishing important environmental movements in recent decades. There was Rachel Carson's clarion call to curb pesticide use, women's grassroots activism to end toxic pollution in Love Canal and other local communities, and women's activism in the European Green party. Yet today women have been excluded from leadership positions in environmental organizations in the North. Moreover, these organizations have shifted toward hierarchical and corporate organizational styles that reproduce the masculine cultures responsible for environmental destruction in the first place. The move toward these kinds of masculine organizational styles that stress the importance of expertise over "merely social" considerations can be seen more broadly in the redefining of environmental issues as scientific rather than social and political ones. Such shifts away from women's (and other cultures') standpoints on nature and social relations are obstacles to developing sustainable environments according to many reports.[21]

Insofar as gender relations structure social relations, human interactions with nature, and their meanings, there will be distinctive women's standpoints on human environments that both enable and limit human knowledge about and interactions with nature. Theories of nature, science, and history that ignore or devalue these reproduce ignorance and error and disable the projects that draw on such theories. Of course it is not just the northern post-Kuhnian and southern postcolonial prefeminist science studies that have done so; with important exceptions, northern feminist science studies have not yet been able to center the multicultural and global feminist standpoints that have proved so valuable in other disciplines and policy areas.

The next chapter turns to examine the important resources for such a project that are provided by the "women and alternatives to development" projects. It looks also at the challenge they raise to the modern versus premodern conceptual framework that has situated European sciences and technologies of the last five centuries in a different category of knowledge production from other cultures' scientific and technological traditions. Further, it looks at the importance of women's full participation in designing more effective "development" policy for both their communities' well-being and the growth of more democratic forms of social progress in general.

7

Gender, Modernity, Knowledge: Postcolonial Standpoints

1. A Problem. Is "social progress for humanity" also progress for women? Or even for all men? Historian Joan Kelly-Gadol raised the issue about Renaissance women more than two decades ago. She went on to argue that this situation evidently was not unique for women:

> The moment . . . one assumes that women are a part of humanity in the fullest sense, the period or set of events with which we deal takes on a wholly different character or meaning from the normally accepted one. Indeed, what emerges is a fairly regular pattern of relative loss of status for women precisely in those periods of so-called progressive change. . . . [I]f we apply Fourier's famous dictum—that the emancipation of women is an index of the general emancipation of an age—our notions of so-called progressive developments, such as classical Athenian civilization, the Renaissance, and the French Revolution, undergo a startling re-evaluation. For women, "progress" in Athens meant concubinage and confinement of citizen wives in the gynecaeum. In Renaissance Europe it meant domestication of the bourgeois wife and escalation of witchcraft persecution, which crossed class lines. And the Revolution expressly excluded women from its liberty, equality, and "fraternity." Suddenly we see these ages with a new, double vision—and each eye sees a different picture.[1]

Similar questions have been raised from postcolonial standpoints about the voyages of discovery, European expansion, the emergence of modern sciences and technologies, and other such processes regarded as indications of the progressiveness of European society, as earlier chapters showed.

What about modernity itself? To be modern is regarded in many parts of the world as a distinctive mark of social, political, and intellectual progress. Yet skeptics have pointed out that not everyone benefits from the emergence of modern social, political, economic, and intellectual structures, and that the costs of such social changes are rarely fully acknowledged—or even detected—by those who benefit from them. In particular, women's knowledge, in both northern and southern cultures, is invariably conceptualized as premodern and therefore not socially progressive. It is represented as a kind of folk belief, merely local knowledge, or ethnoscience. Such local knowledge is either to be replaced by modern scientific knowledge or tolerated as a kind of secondary, intuitive, auxiliary, or applied knowledge, not to be valued as highly as modern scientific knowledge. In such contexts, modernity is taken to be identical to European modernity; hence, it is European culture that provides the distinctive mark of social, political, and intellectual progress. Should it?

What is meant by "women's knowledge"? The last chapter explored a conceptual framework that can be useful for understanding that and how women, like men, have distinctive relationships to the nature and social relations around them. Northern feminist science and technology studies have provided plenty of examples of women's distinctive patterns of knowledge. As professional healers, midwives, and nurses, women have been repositories for and developers of knowledge about everyone else's bodies. Women's knowledge of our own bodies often has proved more reliable than modern biomedicine's diagnoses. Women today, as in the past, perform much of the daily care of sick and aging relatives and develop distinctive patterns of knowledge that often enable them to diagnose what ails their elderly kin more accurately than can physicians and other health care workers. These kinds of activities have also required knowledge of pharmacological and other therapies.[2]

Moreover, women are assigned mothering responsibilities. Whosoever does mothering activities, be they women or men, must develop distinctive patterns of knowledge about the natural and social worlds of infants and children in order to address the challenges such responsibilities entail. Women develop distinctive repositories of emotional knowledge, initially through early childhood development processes and in adult life through the assignment to them by gender roles of emotional work in families, workplaces, and communities. Such resources lead to the emergence in girls and women of distinctive patterns of moral and epistemic reasoning.[3] And this kind of list could continue, for other research projects have shown how women similarly develop and maintain distinctive patterns of systematic knowledge in their wage labor, community activities, and other interactions with natural and social environments. These are kinds of knowledge usually represented as outside the sciences, and it is this women's knowledge thought of as "outside science" that will be the focus here. (The last chapter gave examples of women scientists' distinctive patterns of knowledge and ways of producing it in primatology, biology, and other scientific fields.)

There is a more accurate, revealing, and useful way to understand this women's knowledge than the typical way it is presented in prefeminist and sometimes feminist writings. To begin with, one can summarize the origins of such knowledge patterns in the following way: insofar as women and men interact with different regularities of natural and social worlds, have distinctive interests in those regularities and in others that they share, stand in different relations to available discursive resources (metaphors, models, narratives, etc.), tend to organize differently the production of knowledge, and occupy a distinctive location in their culture's diverse and complex power relations, they will tend to produce and sustain different patterns of knowledge and ignorance. This kind of conceptual framework for thinking about such gendered standpoints on nature was examined in the last chapter, where it was presented as useful for thinking about women's knowledge in any culture, and proposed for accounts of local knowledge in general in chapter 4.[4]

The examples cited above are familiar from northern feminist writings.[5] However, the debates focussed on postcolonial, southern concerns about women, the environment, and alternatives to development (WEAD) expand both the domain and force of such arguments. They bring into focus additional patterns of women's distinctive knowledge, for the maintenance of everyday life for which women are assigned a primary responsibility offers different knowledge possibilities in so-called developing and developed cultures. Moreover, the WEAD debates can be understood as an important component of the postcolonial science and technology studies that have emerged since World War II and especially during the past several decades. Thus, they enlarge and deepen the possibilities for dialogue between, on the one hand, northern science and technology policy that often assumes much of the older idealizations of modern scientific processes, and, on the other hand, the northern feminist and southern postcolonial approaches, with both of which they have affinities; they are a site of such dialogue. Additionally, the WEAD debates also offer clues for linking issues about women's knowledge more firmly into current post-Kuhnian "post-empiricist" science and technology studies. Further, they make possible a critical focus on mechanisms that legitimate European modernity. Thus, these debates lie at the juncture of central concerns in diverse schools of post–World War II science and technology scholarly studies and policy agendas. Last but not least, they move concerns of the world's economically and politically most vulnerable peoples—the poorest of the poor who are the majority of the globe's citizens—to the center of discussions about the nature and goals of scientific and technological change past, present and future, and how the social progress that modern sciences are supposed to advance can be understood.

A brief chapter such as this one could not possibly fully demonstrate the richness and complexity of the WEAD debates, or their important convergences and divergences with these other schools of post–World War II science and technology studies. But it can at least try to outline some of the central themes in what such a project would look like. It will do so by focussing only on one issue, namely whether women's knowledge is usefully conceptualized as premodern, as outside modern knowledge systems. Instead, we can see that it is continually developed and maintained as part of the latter. To put the point another way, the WEAD analyses can highlight how "modernity" continually produces but devalues and obscures the importance of "premodernity" in its social systems and philosophies as a condition of recognition of its own "modern" successes. The next section outlines arguments first from the development debates and then from the WEAD contributions to them. The final section identifies some of the philosophical and political implications of such arguments.

2. *Women, environment, and alternatives to development debates.* The WEAD debates arose within discussions about how so-called development was in fact creating a great deal of de-development and maldevelopment rather than only the improvement in the quality of life in the developing societies that had been the ostensible goal of such programs. WEAD emerged in the late 1980s as the latest stage in feminist participation in these globally appearing critiques of 1950s' conceptions of development and their effects, though some of its themes are much older.

Development or maldevelopment? By the early 1970s, the unexpected effects of development policies instituted during the 1950s and 1960s had already been widely identified: development was having more bad than good effects on vast numbers of the world's poorest peoples. Those most in need of development were bypassed and, worse, further disadvantaged by development policies and practices. Development had been conceptualized as economic growth during the early days of the Cold War and the beginnings of the end of formal European colonial rule. Such growth was to be created by the transfer from the United States, Europe, and the Soviet block to "unaligned" and "underdeveloped" societies of patterns of industrialization that were thought to have so successfully modernized Europe and North America from the late eighteenth through the early twentieth centuries. Thus, the scientific methods, information, and attitudes claimed to direct European/American industrialization, as these were understood in the post-war northern societies, were to be central to improving conditions for the poor in the South. Consequently, the appropriateness and adequacy of these methods, information, attitudes, and, eventually, the whole "scientific worldview" became implicated in the failures of development policies.[6] In retrospect one can see that from the beginning trouble was brewing for the conventional assumptions of empiricist science since many of these critics, from the international development agencies no less than from activist groups in the underdeveloped societies, understood perfectly well but did not believe the conventional wisdom that "pure science" could be immunized against such criticisms of what empiricists would insist were merely science's applications and technologies.[7]

Every part of this initial way of conceptualizing what constitutes development was subsequently identified as contributing to the de-development and maldevelopment of southern societies. Two elements were most problematic. One was the equation of development with economic growth. The other was the related assumption that the scientific and technological routes to improved quality of life for the underdeveloped societies in the late-twentieth-century southern hemisphere would be those that improved the quality of life in eighteenth- and nineteenth-century Europe and the United States (bracketing for the moment issues about for whom it improved life conditions and what should count as improvement there and then).

Policies to promote continued economic growth, further analysis revealed, cannot possibly avoid destroying everyone's environments and the cultures and peoples that depend upon them for survival. In the long (and very often shorter) run, economic growth inevitably destroys its own material, environmental base, namely the natural resources needed for even a sustained economy, quite apart from a growing one. It also—through the destruction of environments and through other processes—destroys the cultures and lives of vast proportions of the world's already most economically and politically vulnerable peoples who depend upon local environments and their own local social relations for survival. Development projects have generated inequality between generations as present generations use up the resources necessary for the survival of future generations. They have increased inequality between the "haves" and the "have nots" both within the southern societies and in the global economic system. The "haves," both locally and internationally, seemed to be the ones mainly benefitting from development since they already had the economic, social, and political resources to direct such projects to serve their own interests, desires, and assumptions about what constituted development and what role they should play in such processes. Their benefits have been gained at the expense of the most vulnerable peoples and their cultures. The problem was not so much a matter of rampant individual entrepreneurialism, though that was certainly visible. Rather, multinational corporations, their local managers and allies, and national and international northern-controlled government agencies were most able to muster the resources that enabled them to direct development projects in ways that responded to their perceptions of what development was and should be about. These perceptions usually had been shaped by conceptual frameworks developed in the context of older eurocentric, capitalist, racist, and colonial projects.

Additional destructive effects of development were produced through the North's proliferation of militarism both within and targeted at developing societies. Of course Cold War politics infused development policies with military components. However, militarism was independently the expectable companion of northern development policies that from their beginnings were conceptualized and functioned as a continuation of half a millennium of European (and more recently U.S.) expansion. For example, the conquistadors and the local skirmishes they aided and abetted in the Americas came to be understood as the predecessors of late-twentieth-century military engagements generated at least in part from the North that were located within and aimed at peoples in developing societies.

Of course some development projects had at least some of the good effects intended on at least some people. For example, the green revolution did indeed raise food availability significantly for large numbers of people. However, the costs of such good effects were that it also increased the control of northern agribusiness over Third World farming, further supported class disparities within the southern societies, was apparently incapable of bringing many millions of the world's poorest children and adults out of malnutrition and starvation, and even worsened the poverty of the poorest poor. And it destroyed or used up local environmental resources at accelerated rates. Many other projects could not even exhibit the virtues of the green revolution record, for they had amounted to little more than "colonization by other means," according to observers from the North as well as the South.

Hence, the definition of development as economic growth was one conceptual issue implicated in development failures. The second, related one, was the enshrinement in development policy thinking of the model of eighteenth- and nineteenth-century European and U.S. patterns of scientific and technological change (as 1950s' thinking understood them) as universally valuable for every society's economic, social, and intellectual progress, in any part of the natural and social world, in any historical era. As a subsequent generation of northern post-positivist, post-Kuhnian histories, sociologies, ethnographies, and philosophies of science have analyzed in great detail, this empiricist or positivist representation of scientific and technological change—the internalist model is neither empirically nor theoretically adequate to explain even the history of modern scientific and technological change in Europe for the last five centuries. Such change cannot be isolated from its local sociopolitical formations in the way the internalist model leads us to believe. Instead, the social studies of science and technology have shown how sociopolitical formations and their favored conceptual frameworks for producing knowledge co-evolve, emerging together and continually interacting during processes of historical change. Most relevant to the WEAD debates, modern sciences have not been unified in the strong senses of this term assumed by the internalist model. Instead, modern sciences' integration with diverse social, political, moral, intellectual, and technological formations of their own historical eras has insured their great successes as well as their limitations, and in either case their disunity. This history was discussed in earlier chapters and will be pursued further in later ones. The last section returns to characterize further some of the central tendencies in this northern literature that the WEAD debates can usefully inform.[8]

Moreover, the eurocentric dimensions of the conception of scientific and technological change directing development policies began to be revealed by postcolonial histories of science and technology, as earlier chapters recounted. These began to appear at the same time but in only sporadic interaction with the northern post-Kuhnian studies. According to these accounts, modern sciences had emerged in Europe in part because of the resources provided to them by the Voyages of Discovery and subsequent stages of European expansion, just as European expansion was made possible by the resources modern sciences provided. For much of scientific development and for all of European expansion, neither could have occurred without the other. Crucial to the advance of both European expansion and of modern science was the European appropriation in some respects and destruction in others of the scientific and technological traditions of non-European cultures. Moreover, according to these accounts, post–World War II "development" amounted primarily to more of the same; it was simply European expansion in sheep's clothing.

Feminist analyses. This was the context within which the WEAD debates emerged. They began with criticisms that women were being left out of development. Only men were being trained in the agricultural, medical, manufacturing, and other scientific knowledge and technologies that were to improve the quality of life in so-called underdeveloped societies. From this early "women and development" or "women in development" phase, the gender analyses moved on to identify problems with how development was being conceptualized and practiced. It was not just that women were bypassed by development policies; even worse, sexist and androcentric northern assumptions in such policies had the effect of de-developing and maldeveloping women. Such worsening conditions for women then deteriorated conditions for their children, households, kin, the elderly and sick, and the communities that women's resources had enabled to thrive. It became clear that when women were de-developed, so too were their societies.[9]

For example, under development policies, not only were women excluded from opportunity to get the scientific and technological training to improve their agricultural projects, but also their traditional land rights tended to shift to the men who were the only ones being taught scientific farming. This loss of traditional rights impoverished women's kin and dependents also. Moreover, often scientific irrigation and energy projects left women with increased workdays necessary to gather water and fuel and accompanying declines in their nutrition and general health and that of their dependents. Sometimes, they were left sole supporters of their families and communities as their fathers, husbands, brothers, and sons were attracted to the new industries in the cities or to other wage work far from home, such as mining, manufacturing, or agribusiness. Pesticides damaged the health of women engaged in the new scientific agricultural production. As northern agribusiness continued expanding into these societies, women in rural areas were no longer able to support themselves or their dependents. Their alternatives were domestic work in cities or abroad, prostitution in the tourist industry or around military bases, or exploitative manufacturing work in the "free trade zones," all of which left them unprotected by their traditional guardians—family and local governments—as they were in effect pimped by both to northern projects and the local allies they promised to enrich and empower. Recently, the demands of northern financial institutions that their development loans be repaid by cuts in social services—literacy programs, nutrition supports, health care, child care, and the like—has disproportionately shifted to women the burdens of paying for these development projects. Such patterns reveal that the northern financial institutions and their beneficiaries have been most "developed" by loans to the South.

As testimony to the failures of development mounted and feminist analyses in general became more sophisticated in the diverse disciplines contributing to development thinking, yet another important feminist insight emerged. It became clear that far from being left out of development or maldeveloped only as an unintended byproduct of development, women and their resources were centrally involved in prevailing development policies and practices, though not in ways that benefitted women. Appropriating women's resources—land, time, food and energy resources, paid and unpaid labor, community networks for maintaining health, political and economic "insurance," and so on—was central to economistic development's most exploitative but "successful" practices. It has been the appropriation of women's land and land rights, women's underpaid wage labor, and women's unpaid reproductive and productive labor that have provided some of the most significant sources of economic "growth" for their societies.[10] To put the point another way, it is the appropriation into European economic forms and the derailing of their ongoing activities there (including knowledge production) that have enabled the successes of modern economistic "development." Alongside the mainstream criticisms of economistic development emerged a feminist call for a different model—an alternative to "development"—to avoid such processes that increase exploitation of the already most economically and politically vulnerable in ways that benefit most the already most advantaged. More accurate assumptions from recent feminist and environmental thinking must be used to redirect development planning, the WEAD theorists argued.

"Development is gendered": Women's concerns are fundamental to success at improving the quality of life. A good example of an analysis assessing such prospects is provided in Wendy Harcourt's "Negotiating Positions in the Sustainable Development Debate: Situating the Feminist Perspective."[11] Of course, no one author or essay can represent the diversity of a complex and globally developed approach such as the WEAD analyses, and the report here even of just this one will be brief. However, the fact that this analysis assesses the strengths and limitations for addressing women's issues of the best of the criticisms of economistic development policy makes this essay especially useful for grasping at least the outlines of both the political and intellectual strengths of the WEAD approach.

Sustainable development has become the goal of a wide array of policy theorists critical of economistic definitions of development. They propose "a new productive ethic at the heart of development which will value the quality of peoples' relations with each other and with nature" (12). Thus, sustainable development is understood to require equity between generations, in the sense that the needs of future generations are equitably planned, and a balance between economic, social, and environmental needs. To attain such a balance, the conservation of non-renewable resources and a decrease in the economic and social costs of the wastes and pollution produced by industrialization must be achieved (1–2).

As Harcourt notes, to set sustainable development as a goal certainly is an improvement over the earlier policies that sought only economic growth. However, she points out how the goals of the prefeminist sustainable development theorists cannot be achieved until the insights of feminist and environmental thinking are used—or used in more sophisticated ways—to revise even these valuable corrections to econ-omism. To start off, the purportedly gender-neutral selection of development goals that the sustainable development theorists propose, described above, does not well represent women's concerns. In this aspect of social relations as in others, the assumption or advocacy of gender neutrality marks positions that cannot avoid advantaging men since the distinctively human (or modern, or Chinese, or moral, or historical, or philosophical, etc.) systematically has been constituted in terms of the manly, and vice versa. Some of the development theorists apparently think that successful sustainable development policies and practices can ignore feminist concerns when addressing economic and environmental concerns. Yet these concerns are in fact interdependent. Women's well-being depends on sustainable environments and viable non-exploitative economic relations. Such economic relations depend on women's well-being and healthy environments. And healthy environments depend on viable non-exploitative economic relations, including with women, and on the resources that women, too, can bring to interactions with environments.

Harcourt identifies four schools of thought about sustainable development that share the view that economic growth causes destruction of the very environment and social relations necessary to sustain people's lives, not to mention to improve them. Each of these approaches in different ways offers more resources for addressing women's and environmental concerns than did narrow economistic development thinking. Nevertheless, the ways the first three of these approaches conceptualize their projects only goes part of the way toward addressing the concerns of women and environmentalists. And the fourth approach can be expanded in illuminating ways. Identifying some of the strengths and limitations of these approaches can quickly reveal some of the distinctive concerns of WEAD analyses.

The first of these, the "real-life economists," do center environmental concerns in their analyses.[12] But their sole "revised ecological economic version of a universal economic model" continues to legitimate only eurocentric and androcentric understandings of economic and environmental realities (14).

> ... [R]ecent feminist scholarship is passed over and gender is ignored as one of the axioms of economic behavior. In this they inherit the gender blindness of traditional economics. But in a study of the social, ecological and ethical dimensions of real-life processes a consideration of gender relations is essential. (14)

To take just one example, while these economists challenge the philosophy of individualism, they cannot detect that "the notion of freedom of choice is not gender neutral" (14). They disable their own critiques in that they cannot identify the dimensions of this philosophy that feminist philosophers and political theorists have shown should be attributed to gender relations. Nor, one can observe, do they grasp how this notion is embedded in racist and eurocentric conceptual frameworks.

The "people-centered economists" are indeed concerned to eradicate poverty, not just to increase economic growth.[13] Their "focus on local-scale development opens up space for women's productive and reproductive activities to be valued," and their vision "suggests some useful strategies for feminists interested in negotiating better positions for women in the sustainable development debate"(15).Yet these critics retain disabling features of the economistic framework both in their language about the importance of "investment in people"and in their unexamined eurocentric position that a universal model of development constructed with northern assumptions about "equity, resources, household and informal economy" can be adequate for southern societies where such phenomena work differently (15).

The power/knowledge analyses of development make good use of feminist analyses, and they center women's concerns.[14] Yet they fail to locate negotiating positions with other parties to the development debates. Harcourt argues that they falsely situate themselves as entirely outside development discourses in which they are in fact players. Such a misrepresentation of their own status makes them far less effective than they could be. Moreover, their overly pessimistic position on the possibility of any process conceptualizable as development inappropriately locates no positive role for the West, for industrialization, or for modernity (17).

Finally, especially valuable, Harcourt argues, are the "culture, economics and modernization" theorists, who bring useful resources to the project of correcting the flaws of the other critiques in their re-evaluation of the role of culture and development in relation to modernization projects.[15]

> They ask whether development discourse should be so confidently promoting the Western model of production—where the productivity of the work is perceived as satisfying individual desires rather than fulfilling community needs—as a universal model which is the best option for the organic growth of non-Western cultures. (18)

They also question the western assumption that modern science, grounded in logic and rationality and isolated from the embodied knowledge that develops through daily activities, is in fact the uniquely desirable universal form of knowledge. This is the model shaping economistic development policies, but it cannot avoid the bad consequences that in fact it has produced. Such a model devalues and legitimates the destruction of the forms of practical knowledge that have developed within communities over time and that enable the sustainability of environments, effective but non-exploitative economies, and valued social relations. Women's activities and other cultures' local knowledge and productive systems cannot be valued when western models of modernization are identified with social progress (18–20). It must be noted that these theorists do not idealize traditional cultures, nor do they assume that modernization discourses can have no positive role to play. For example, they point out that modernization can indeed be a liberating discourse in challenging forms of fundamentalist or other "traditional" cultures of sexism.

Harcourt points out that this approach can be the source of a radically different theory of the limitations of the modern, scientific epistemology that has shaped western development models and of the resources available for other, more accurate and democracy-advancing accounts of how knowledge of natural and social worlds does and could develop.

> For a feminist critique these writings based on observations of Third World cultures and perceptions raise the question of whether modern economics might not learn from other ways of social and economic organization, rather than always assuming a dominance and superior knowledge. Feminist critics place development economics, with its devaluation of nature and its failure to treat other cultures and people with dignity, as just one system of knowledge in the process of modernization, which needs to be examined critically for its continuing worth, given its current poor record at solving the problems it defines and sets out to solve in development. In doing so these writings give economics a cultural context which helps deconstruct its claims to objectivity and truth. (20)

We shall return to this vision of the reality and value of a disunity and multiplicity of sciences in a later chapter—a theme that has emerged from all three schools of post–World War II science and technology studies.[16]

All four of these lines of analysis are beginning to shape national and international development policy, though the first two more powerfully than the last two. After all, the legitimacy of the economistic model of development faces overwhelming public skepticism around the globe because of the now inescapable evidence of how it continues to destroy southern economies, environments, and cultures, further disadvantaging the already most economically and politically vulnerable peoples as it distributes benefits primarily to the most advantaged peoples in the North and their allies within the southern cultures. Thus, Harcourt usefully argues that the challenge for WEAD theorists is how to form alliances with these existing progressive development discourses while simultaneously getting feminist issues centered in them.

A crucial strategy will be to demonstrate the centrality of women's concerns to the success of the progressive development programs. Drawing from Harcourt's and other feminist development analyses cited earlier, one can identify seven points that would contribute to such an argument. First, as indicated earlier, women's resources sustain everyone in their communities. To de-develop or maldevelop women is to do so also to all of those dependent on women for daily survival:

children, other household members, kin networks, the elderly and sick, and the larger communities that women's activities sustain. The question is then whether women's resources, including their distinctive patterns of knowledge of natural and social relations, will be permitted to sustain women and their own communities or be appropriated instead for the benefit of the already most-advantaged in the North and South.

Second, women sustain healthy local environments. They are not "closer to nature." However, women in many developing cultures are assigned maintenance of the local environments that produce food, shelter, fuel, water, and other necessities on a daily basis. As members of communities, not as individuals, they are repositories for and systematic developers of such local knowledge. To de-develop women is to deteriorate the resources communities have for sustaining their environments. Moreover, "local" environments cannot be bounded by political treaties. They are interconnected by nature's laws and social practices; thus, what happens to one culture's environment has consequences for others.

Third, women's de- and maldevelopment increase population growth. Women's poverty and lack of education, which reduces their opportunities for income producing activities, make childbearing the most reasonable economic strategy for them and for their families. Women's poverty and illiteracy is the major cause of overpopulation and accompanying environmental degradation—as even the U.N.'s 1994 Cairo population conference finally officially recognized. Thus, insuring women's economic, political, and social welfare is the most potent antidote to overpopulation.

Fourth, women have fewer interests in militarism. Militarism is an issue for women in developing countries since, for one, some of their countries have been the open targets of northern militarism, and many have suffered from the increases in military aid that arrived as part of "development" efforts to recruit allies to the support of Cold War politics. Moreover, local resources devoted to militarism, whatever the origins of the latter, take resources away from other projects. It would be historically inaccurate to attribute essentially peaceful natures to women and war-loving ones to men since women have happily gone to battle and vociferously agitated for war, and men have been conscientious objectors and have organized peace movements. Nevertheless, objectively it is clear that at least in the late twentieth century, around the globe women and their children have died in greater numbers and suffered more from warfare than have men, and they gain fewer symbolic or material benefits from militarism. Moving women to central positions in national and international political and economic circles carries the possibility of defusing militarist tendencies.[17]

Fifth, women's assigned social roles position them as powerful change agents. Women educate future generations. Women's responsibilities for child care and their management of community resources—health, food supplies, kin networks—position them as central resources for changing present and future attitudes and practices.

Sixth, women have leadership, negotiation, and policy making skills that are crucial to processes of social change. They are used to working together, and to negotiating with other women and with men. They can less often resort to violent acts to achieve their social goals. They have the social skills necessary to work in and lead the public policy making from which they conventionally have been systematically excluded.

Last, one can find in the writings of some feminist critics of development policies the argument that to women's concerns, skills, and knowledge already should be attributed much of whatever counts as the successes of economistic development, though such contributions are suppressed in the economistic accounts. Such successes could not have occurred without the continued appropriation, through economic policies that have disproportionately benefitted advantageously positioned groups, of women's and peasants' lands, labor, and knowledge. Such "economic growth" has been at least in part on the backs of women and peasants. And this insight leads to rethinking how women's material resources, labor, and knowledge, unacknowledged or devalued though they be, have provided valuable contributions to advancing "modernity" itself.

3. Philosophies of gender, modernity, knowledge. As a preamble to considering the modernity issue, one can note that the WEAD analyses produce conceptions of gender, scientific and technological change, nature, and social progress that can be valuable resources for the other post–World War II schools of science and technology studies and their philosophies. It is not that these southern feminist science and technology writers engaged in the WEAD discussions are the only ones ever to have arrived at their insights. (Of course, we could decide by definition that all accounts that met the adequacy standards of the WEAD accounts were thereby "southern accounts.") Rather, their arguments emphasize and expand the importance of centering the conceptions they have found useful.

Of course Rational Man has disappeared from these accounts as a possible or desirable agent of history and knowledge projects. He was present in the economistic conceptions of development, making his rational economic choices that resulted, it turned out, in de-development and maldevelopment. But so, too, have disappeared attempts to conceptualize proposed agents of social progress in terms of any other universal class, such as the proletariat or "woman." Gender relations obviously are constructed in culturally specific ways in the diverse populations that are to help design and benefit from "alternatives to development." Consequently, it is easier to resist the essentialist tendencies to assume there is some common characteristic that all women share if one starts off thinking about scientific, technological, and epistemological issues from the standpoint of the lives of the specific groups of women reported in the WEAD analyses.[18] In these WEAD accounts, gender is understood as culturally specific structural and symbolic relations, rather than only as a property of individuals' identities.[19] The important point here is that the configurations of gender relations in any particular culture cannot be known prior to inspection. A continuing project in the gender and sustainable development studies is to chart how structural and symbolic gender relations are manipulated by international and local elites to their benefit, and how they change over time in response to other changes, including scientific and technological ones. Such respect for the integrity of local social formations counters the conventional tendency to view gender as fixed by nature, tradition, or socialization processes, and as fundamentally a property of individuals.

In the WEAD accounts, nature and scientific change "that works" are also conceptualized differently than in conventional northern philosophies. It must be noted that sometimes "science" is used in these development discussions to refer only to northern modern sciences that are being distinguished from other cultures' systematic knowledge about natural and social worlds, and sometimes it is used more generically to refer to any culture's such systematic knowledge. The focus here is on the latter, more inclusive sense in which the term is used.[20] Scientific change is always perceived as far more than mere intellectual change, and far more than the changes in the "high sciences" of physics, chemistry, and the more theoretical reaches of biology upon which northern science studies has tended to focus, for both useful and problematic reasons.[21] The production of systematic knowledge about natural and social worlds is treated as a co-evolving part of larger social and political formations, each contributing to the historical trajectory of the others.[22] Thus, women's local knowledge in non-western societies develops alongside of modern sciences when the latter is introduced—a point to which we shortly return.

Consequently, to understand "nature's (local) order" and how to intervene in it effectively, one must also understand local social orders and how to negotiate within them. Sciences are never pure, but always a mix of natural and social knowledge, whether their explicit topics fall within conventional western categories of nature or culture. Obviously this is so for the environmental, health, industrial, and such mixed social/natural and science/technology knowledge projects that are of most concern in the developing societies. But it is also true of such more abstract disciplines of physics, chemistry, and even mathematics.[23] In these analyses a purely physical nature cannot be isolated from the social relations that shape how cultures conceptualize and interact with their surroundings. Environments are created and maintained by histories of interactions between cultures and their surroundings. In the WEAD accounts, the "nature" that modern European science observes does not appear as an object of universal human experience that can be abstracted from its local conditions of conceptualization in equally useful ways for every culture at every historical moment. Rather it appears as just one among many notions distinctive to their cultural context that have been useful for thinking about the environments with which humans interact and that they are part of, as the importance of features of such environments are selected by local interests, discursive resources, and culturally typical ways of organizing the production of knowledge, whether or not these involve scientific methods.

Finally, in these discourses the debates over how development is to be defined are debates over how social progress is to be conceptualized and measured. A radical democratic conception of development is adopted which substitutes for the goal of advancing the well-being of Rational Men that of advancing the well-being of environments and cultures upon which people depend for survival, including women and the rest of the world's most economically, politically, and socially vulnerable peoples.

What kinds of philosophies of modernity and knowledge are called for by these aspects of the WEAD debates—ones that would assume such conceptions of gender, nature, scientific and technological change, and social progress? Some clues to distinctive features of such philosophies can at least be identified here. First, epistemology and philosophy of science would always be recognized to have political dimensions. They have political origins, processes, and consequences since who does and who does not gain access to defining, making, legitimating, and deploying a culture's knowledge systems is a political matter. Power, including gender advantages, is accumulated, deployed, and distributed through epistemological and philosophic processes.

Second, such epistemologies and philosophies of science would have to be informed by the postcolonial histories that charted the co-constitution of European sciences with the global social relations of their successive eras. Feminist accounts, too, must be informed by such studies. Global economic, political, social, and psychic formations have shaped modern sciences' distinctive patterns of systematic knowledge and systematic ignorance from its very beginnings five centuries ago (and earlier) through its interactions in the post–World War II gendered global political economy, just as modern sciences have significantly helped to constitute economic, political, social, and psychic formations of their historical era. Gender relations have played a significant role in the emergence and maintenance of what counts as distinctive modern sciences and their patterns of knowledge and ignorance. And such accounts of modern sciences as local European knowledge systems should not be limited to the perspective on such histories provided by northern natives.

Thus, third, other cultures' local knowledge systems will also be conceptualized as dynamic constituent parts of their social formations, changing over time as they interact with European social formations, as well as others. For example, the Voyages of Discovery no less than post–World War II development policies have brought European and other local knowledge systems into diverse interactions with each other, changing all through such processes. A single historical and philosophic conceptual framework can be useful to represent how both modern sciences and other local knowledge systems come into existence, change over time, and in some cases, disappear. "Women's knowledge systems" in their diverse local forms would then appear as just one component of more and less global knowledge systems, continually changing as they interact with knowledge systems originating in other culturally distinctive locations in nature, interests, discursive resources, and ways of organizing the production of knowledge. They would appear as co-produced within other knowledge systems, some aspects of which they complement and others with which they compete, as culturally distinctive locations in nature, interests, discursive resources, and ways of organizing the production of knowledge lead would-be knowers in converging, diverging, and oppositional directions.

Fourth, explanations of the successes of modern sciences could thus not be restricted to accounts of their unique or internal epistemic features, such as their standards of objectivity and rationality, their distinctive method, critical attitude, reliance on formal languages, distinctive ontologies, or the like, though such "local" features do have historically specific roles to play in enabling and limiting the growth of knowledge in modern sciences. Such purportedly "internal" features are always linked to larger social formations and their gender relations. The unity of science and its unique universal validity no longer appear plausible from the perspective of the postcolonial WEAD debates.

Fifth, within the conceptual framework produced by the WEAD studies, many current heated controversies become topics for historical and philosophic analysis rather than, as they are conventionally represented, the questions requiring choices between the alternatives offered. Here WEAD discussions widen and deepen such tendencies in science and technology thinking more generally. We can ask about proffered choices between realism versus constructivism, absolute (universally valid) epistemologies versus relativist ones, rationality versus irrationality, objectivism versus subjectivism, materialism versus idealism, the natural versus the social, and the individual versus society how they function primarily to maintain androcentric and eurocentric modernist frameworks for philosophies and histories of science, blocking our ability to see other choices beyond these limited ones.

Sixth, the WEAD discussions show the centrality of gender relations not just when "women's issues" are the topic, but perhaps even more significantly, when they are not. WEAD analyses provide additional evidence for the argument that claims to cultural neutrality are themselves always culturally local. Development projects are themselves gendered masculine and marked as eurocentric to the extent that they ignore or do not center the concerns of any but the institutions designed to serve the economically and politically most advantaged, a group from which women and the poor, including vast majorities of southerners, have been excluded. Insofar as modern scientific projects fail to achieve recognition of their gendered locations on postcolonial maps of natural and social geographical locations, interests, discourses, and ways of organizing the production of and standards for knowledge, they remain marked by their containment within androcentric and eurocentric perspectives.

Finally, women's knowledge, in the North or South, could be reconceptualized outside the modern-versus-premodern dichotomy. Local knowledge, whether here or there, is not reasonably understood as either merely a precursor to future universally valid scientific reasoning or as merely applied science. We saw in earlier chapters that all knowledge systems are local ones in important respects. We saw here that women's knowledge has been and remains crucial to the advance of "modern" knowledge. Thus, women's knowledge is just one necessarily continuously produced and reproduced element of global systems of (always only local) knowledge. In such global systems, modern sciences require the continued production of other local systems of knowledge that can be appropriated into modern sciences, that can evaluate which parts of modern science are valuable and which are not for specific local projects, that can figure out how to apply modern scientific principles and information in local and changing natural and social contexts, that can identify elements of other local knowledge systems with which modern scientific elements can be integrated or combined, and so on. Such other knowledge systems also are dynamic ones, continually adapting to the resources and problems created by modern sciences and their social formations, and by constantly changing natural worlds, among other change agents.

Is "social progress for humanity" also progress for women? Not now. However, social progress for women is always also progress for humanity, as the WEAD discussions have highlighted. Thus, "social progress for humanity" will also deliver progress for women only when the knowledge of women and other such groups that provide crucial resources for "modernity" are more realistically evaluated. Such a project requires many kinds of social transformation. One such transformation will be in the epistemologies regarded as most useful and compelling.[24]

8

Recovering Epistemological Resources: Strong Objectivity

What kinds of theories of knowledge will serve as reliable guides in the world revealed through postcolonial, feminist, and post-Kuhnian science and technology studies? The old theories insisted upon the possibility and desirability of a culturally neutral science that was to be insured by its distinctive method; to be exercised in the context only of justification; to produce one uniquely universally valid perfect reflection of nature's order; to be discovered by communities of expert knowers who could be isolated in their scientific work from the social currents flowing through their public (and "private") life. This dream of a single, perfect model of knowledge has been lost forever under the rigorous gaze of the various schools of post–World War II science studies discussed in earlier chapters. Though it no longer represents state-of-the-art history, philosophy, or other approaches to understanding and explaining science and technology, this internalist epistemological vision still exerts a powerful hold on the imaginations of most people outside these disciplines and some who are inside them.[1] This dream remains especially compelling for people in professional and administrative classes around the globe who have received great benefits from modern scientific and technological change. Of course it is hard for anyone to bring the most rigorous critical examination to bear on ideals that so powerfully support the legitimacy of one's favorable circumstances.

But does giving up this dream require rejecting the epistemology of modern science completely? As indicated in earlier chapters, the answer here to this question is a resounding "no." This chapter and the next three focus on how new ways of thinking about epistemological issues that have emerged from post–World War II historical, sociological, ethnographic, and philosophic studies of sciences and technologies can be used to retrieve and make functional again some important insights that stimulated the now archaic "dream." To put the point another way, it is not just these new science and technology studies that possess all the valuable epistemological resources for the worlds of multicultural, postcolonial, global, and new gender relations that lie before us. The European scientific and epistemological legacy also can continue to make valuable contributions to these worlds. This may seem to many readers to be a naive and patronizing comment to make about modern European sciences and epistemologies. However, post–World War II science and technology studies are often understood to produce only a critical posture to the European legacy. My point here is that their critical posture also is accompanied by an appreciation of the power and value of this scientific and epistemological legacy—albeit one that differs in significant respects from conventional appreciations. The analyses here will try to locate epistemological implications of the post–World War II schools of science studies firmly within still valuable aspects of the European legacy. The question should be not how to preserve as if carved in stone or else to completely reject the European legacy, but rather how to update it so that it, like many other "local knowledge systems," can be perceived to provide valuable resources for a world in important respects different from the one for which it was designed.

These analyses can also more clearly sort out just what the epistemic implications are of the new science and technology studies. Are the post–World War II, post-Kuhnian, feminist, and postcolonial accounts of nature and social relations supposed to be "truer" than the familiar, supposedly value- and interest-neutral ones? What happens once cultural elements in sciences and technologies are understood to consist not just of "values and interests," but, more extensively, of culturally distinctive interests, discursive resources, and ways of organizing the production of knowledge?[2] On what grounds can anyone persuasively support one of several competing accounts of nature or of the history of science if there are no longer plausible transcultural standards for justifying our beliefs? Must all knowledge, including of the laws of nature, find support only as a local point of view? If so, is all knowledge, including our best scientific claims, only relatively valid? Are there any uses left for such concepts as objectivity, rationality, good method, and the universal validity of scientific claims? And what about this very account itself—the account in this book? On what grounds should readers find it more satisfactory than its competitors? What are its limitations? What forms of reflexivity fit with the other epistemological concepts developed in these new kinds of science studies?

This chapter focusses on the notion of objectivity, proposing ways to extract it from its contexts and projects that are no longer regarded as viable, and to strengthen it so that it can function effectively in the world created by post–World War II social, economic, and political histories and the science and technology studies with which they have co-evolved.

Let us begin by noting some features of the new context within which any questions about objectivity must interact, whether or not we who ask such questions have such contexts in mind.[3]

1. *Science, society, and the new "objectivity question."*[4] As the last five decades of science and technology studies have made clear, observations are theory laden; our beliefs form a network such that none are in principle immune from revision; and theories remain underdetermined by any possible collection of evidence for them. There are always many additional possibly plausible hypotheses about any state of affairs that have not yet been proposed, or have been considered but perhaps prematurely dismissed, and thus remain untested at any given moment in the history of science. Some small subset of them undoubtedly would fit the existing data just as well as whichever one is favored at present. After all, sciences produce new theories continually. As one analyst has put the point, many scientific theories can be consistent with nature's order, but no one can be uniquely congruent with it.[5] In short, there is enough slack in scientific belief sorting to permit social values and interests, discursive resources, and ways of organizing the production of knowledge not just to leave their traces on otherwise culturally neutral results of research, but to constitute scientific projects in the first place.

This is not to say that science "makes up" its results of research, or that science is "nothing but politics by other means," of course, as familiar internalist epistemological approaches invariably interpret such findings of post–World War II histories, sociologies, ethnographies, and philosophies of science and technology. Rather, scientific and technological projects co-evolve with other elements of their particular historical social formation—new forms of local and global economic relations, of the state, of educational systems, of religious practice, of gender relations, of child-rearing, and so on. The whole social formation, including its scientific and technological projects, is "constrained" by nature's order, as these accounts make perfectly clear.

Most importantly, this co-evolution of sciences and the rest of their social formations turns out not just to limit the growth of knowledge, as it always does in one way or another, but also simultaneously to be a resource for its growth, enabling different cultures, and different historical eras in the same culture, ever to detect yet more aspects of nature's order. Such processes permit more than one theory to fit any set of observations, more than one interpretation of any theory to be reasonable, and, consequently, the growth of science in ever new directions, as two philosophers put the point.[6]

Moreover, modern so-called European scientific and technological projects and those of other cultures, past and present, are in important respects on a continuum and these days usually part of one single global knowledge system.[7] Most obviously since 1492 and the beginnings of five centuries of European Voyages of Discovery, the growth of modern sciences and technologies in Europe and the relative decline of other cultures' knowledge systems have been causally related. So-called development processes since World War II continue to "turn the world into a laboratory for European sciences" in many respects,[8] though at the same time one can see how in many parts of the world, including the North, modern sciences in fact depend upon the continued production and reproduction of other local knowledge systems for their legitimacy and continued growth.[9]

One consequence of such skepticism about the older internalist theories of knowledge appears in the shift from the old to the new objectivity question. The old one asked "Objectivity or relativism? Which side are you on?"[10] The new one takes this question itself to be a topic of discussion—a historic and epistemic problem to be explained. The new one still is directed toward many of the concerns addressed by the older question: Which of the competing grounds for claims about nature and social relations should we prefer? How can we block "might makes right" in the realm of knowledge production? How can we systematically identify widespread cultural assumptions about both nature and social relations, and the social projects that generate them, which have distorted so much of what heretofore has passed as universally valid scientific knowledge? However, the new objectivity question takes the status and underlying assumptions of the old objectivity question also to be one of its problems. It asks, what should be rejected and what saved of the older objectivism?[11] How can the notion of objectivity be updated so that it is more useful for contemporary attempts to understand nature and social relations?

Some readers may well think that now is none too soon to define what objectivity is for the purposes of this discussion. However, this urge should be resisted or, rather, approached in another way. One problem is that the term has no single reference in the prevailing discussions. For example, it has been applied to at least four kinds of entities. Objectivity, or the incapacity for it, has been attributed to individuals or groups of them, as in "women (or feminists, Marxists, environmentalists, blacks, 'Orientals,' welfare recipients, patients, etc.) are more emotional, less impartial, too politically committed, and thus less capable of objective judgments." Second, it has been attributed to knowledge claims, to statements, where it does not seem to add anything to claims to truth or closeness to truth, or to claims to sufficient evidence. Here, an objective claim is simply one that is better supported by evidence—more accurate, closer to the truth—than its competitors. Third, the notion is also attributed to methods or procedures that are thought to be fair: statistical, experimental, or repeated procedures (in the law, ones appealing to precedents) are more objective because they maximize standardization, impersonality, or some other quality assumed to contribute to fairness. Fourth, objectivity is attributed to the structure of certain kinds of knowledge-seeking communities—in historian Thomas Kuhn's account, the kind characteristic of modern natural science.[12] In other accounts, these are specified as communities of experts, or ones that include members of different classes, races, and/or genders (or that exclude them), or that maximize adversarial relations of rigorous criticism of ideas and claims,[13] or that maximize ideal speech conditions, and so on. Though distinct, these different referents of "objective" clearly are not totally independent in people's thinking. Most obviously, the other three should generate results of research that are less false.[14]

But noting these four distinct references for the term is only the beginning of mapping its convoluted outlines. This chapter cannot take space to continue that mapping here, but this project can be pursued through two recent, highly acclaimed histories of the notion. In one of them, Peter Novick shows that objectivity

> is not a single idea, but rather a sprawling collection of assumptions, attitudes, aspirations and antipathies. At best it is what the philosopher W. B. Gallie has called an "essentially contested concept," like "social justice" or "leading a Christian life," the exact meaning of which will always be in dispute.[15]

Some elements in the notion originate in Aristotle's thought; others have arisen in the last few decades. However, "older usages remain powerful"[16] and are called up today whenever people are struggling to determine the place that science, or reason more generally, should have in society. As Robert Proctor, the author of the other history, puts the point about the neutrality ideal that both he and Novick see as historically always required of anything deserving the label "objective," "The ideal of value-neutrality is not a single notion, but has arisen in the course of protracted struggles over the place that science should have in society."[17]

Both Novick and Proctor point out that asserting objectivity sometimes has been used to advance and sometimes to retard the growth of knowledge, and the same can be said of assertions of relativism. Neither position automatically claims the scientific or rational high ground. Nor does either assure the political high ground: each has been used at some times to block social justice and at other times to advance it. As Proctor puts the point, neutrality, the central requirement of the conventional notion, has been used as "myth, mask, shield and sword."[18]

The concerns here are primarily with scientific procedures and methods. Such concerns arise from a widespread observation found in feminist, anti-racist, postcolonial, environmentalist, and other movements for social justice. Such analysts point out that some of the sexist, racist, or other kinds of systematically distorted results of research that have been identified in the natural and social sciences certainly are the consequence of carelessness and inadequate rigor in following existing methods and norms for maximizing objectivity in research practices, as the conventional objectivist accounts always argued. However, a certain range of them is the consequence of something else: of inadequacies in how those methods and norms are conceptualized in the first place. The prevailing standards for good procedures for maximizing objectivity are *too weak* to be able to identify the kinds of culture-wide assumptions that shape the selection of those procedures as good ones in the first place, they argue.

This chapter explores one line of response to the new objectivity question's concern with what can be saved from the older internalist notion of objectivity. That is the program for "strong objectivity" which draws on standpoint epistemologies to provide a kind of method for maximizing our ability to block "might makes right" in the sciences.[19] Maximizing objectivity is not identical to maximizing neutrality, as conventional understandings have assumed. Nor, it can be seen, does it always require it; in a certain range of cases, maximizing neutrality is an obstacle to maximizing objectivity. Though these insights have been developed in the way presented here in feminist theory, they have also been articulated far more broadly. For example, earlier chapters showed this kind of argument widely expressed in the postcolonial science studies.

2. Weak objectivity, or when is neutrality an obstacle to maximizing objectivity? The historical co-evolution of modern ideals of objectivity along with changing modern economic and state forms is described by two well-known biologists in the following way:

> In some ways, the fate of science parallels that of bourgeois democracy: both were born as exuberant forces for liberation against feudalism, but their very successes have turned them into caricatures of their youth. The bold, antiauthoritarian stance of science has become docile acquiescence; the free battle of ideas has given way to a monopoly vested in those who control the resources for research and publication. Free access to scientific information has been diminished by military and commercial secrecy and by the barriers of technical jargon; in the commoditization of science, peer review is replaced by satisfaction of the client as the test of quality. The internal mechanisms for maintaining objectivity are, at their best—in the absence of sycophancy toward those with prestige, professional jealousies, narrow cliques, and national provincialism—able to nullify individual capricious errors and biases, but they reinforce the shared biases of the scientific community. The demand for objectivity, the separation of observation and reporting from the researchers' wishes, which is so essential for the development of science, becomes the demand for separation of thinking from feeling. This promotes moral detachment in scientists which, reinforced by specialization and bureaucratization, allows

them to work on all sorts of dangerous and harmful projects with indifference to the human consequences. The idealized egalitarianism of a community of scholars has shown itself to be a rigid hierarchy of scientific authorities integrated into the general class structure of the society and modeled on the corporation. And where the pursuit of truth has survived, it has become increasingly narrow, revealing a growing contradiction between the sophistication of science in the small within the laboratory and the irrationality of the scientific enterprise as a whole.[20]

Can the kind of revision of philosophic ideals, such as objectivity, play a role in improving this situation? If scientific ideals (and ideas) and social formations co-evolve, as these authors and many others argue, then critically re-evaluating the ideals should be able to make a contribution to critically re-evaluating the discouraging contemporary relations between sciences and society that Levins and Lewontin describe. Ideals can have "material effects," through which they contribute to changing the course of history.

Two politics of science.[21] Has the philosophy of science conceptualized either politics or maximizing objectivity richly enough to meet the widespread criticisms of contemporary sciences and their philosophy that are represented in the passage above? One problem is that the kinds of politics that most threaten the objectivity of science these days escape conceptualization in the conventional internalist philosophies of science.

There are two kinds of politics with which philosophies of science must be concerned. One kind is the older notion of politics as the overt actions and policies intended to advance the interests and agendas of so-called special interest groups. This kind of politics intrudes into "pure science" through consciously chosen and often clearly articulated actions and programs that shape what science gets done, how the results of research are interpreted, and, therefore, scientific as well as popular images of nature and social relations. This kind of politics is conceptualized as acting *on* the sciences from outside, as politicizing a science that was otherwise free of politics—or, at least, of that particular politics.[22] This is the kind of relationship between politics and science against which the ideal of objectivity as neutrality—objectivism—works best, though not perfectly, as Levins and Lewontin point out. It makes sense to think of these interests and values as intruding into science from outside it, and as held by less than (sometimes none of) the group of individuals who constitute legitimate members of the scientific community. In at least many cases, it also is plausible to think of these interests and values as an obstacle to the growth of knowledge. Nazi science, Lysenkoism, or creationist biology are the kinds of examples of such threats to the neutrality of science by political "irrationalism" that the defenders of objectivism have in mind. They do not have in mind the "intrusion" into sciences of forces for maximizing objectivity and enlarging democratic tendencies; any and all "politics" are made to appear equally pernicious to the growth of scientific knowledge in the familiar internalist accounts.

However, sciences are always also shaped by a different kind of politics. Here power is exercised less visibly, less consciously, and *not on but through* the dominant institutional structures, priorities, research strategies, technologies, and languages of the sciences—through the practices and culture that constitute a particular scientific episode. Paradoxically, this kind of politics functions through the depoliticization of science—through the creation of "normal" or "authoritative" science.[23] Thus, a typical standard example that the neutrality enthusiasts cite to demonstrate the bad effects of politicizing science (and they are not wrong about this) can also, paradoxically, be understood as a paradigmatic example of the bad effects of *depoliticizing* science. Robert Proctor describes this example as follows:

> It is certainly true that, in one important sense, the Nazis sought to politicize the sciences. . . . Yet in an important sense the Nazis might indeed be said to have "depoliticized" science (and many other areas of culture). The Nazis depoliticized science by destroying the possibility of political debate and controversy. Authoritarian science based on the "Fuhrer principle" replaced what had been, in the Weimar period, a vigorous spirit of politicized debate in and around the sciences. The Nazis "depoliticized" problems of vital human interest by reducing these to scientific or medical problems, conceived in the narrow, reductionist sense of these terms. The Nazis depoliticized questions of crime, poverty, and sexual or political deviance by casting them in surgical or otherwise medical (and seemingly apolitical) terms. . . . Politics pursued in the name of science or health provided a powerful weapon in the Nazi ideological arsenal.[24]

The institutionalized, normalized politics of eurocentrism and racism, male supremacy, and class exploitation have only here and there been initiated through the kind of violent politics practiced by the Nazis. Nevertheless, they similarly depoliticize northern scientific institutions and practices in ways that shape our images of natural and social worlds and legitimate exploitative policies. Thus, feminist critics have focussed on how gender-coded concepts of the scientist, the "man of professional wisdom," rationality, mechanistic models, "master molecule" models, and so on, escape standard procedures for producing value-neutrality because they have in the first place constituted the scientific institutions and practices that select neutrality-detecting procedures. Androcentric interests, discursive resources, and ways of organizing research constituted modern sciences.[25] And the postcolonial science and technology theorists have pointed out how much of modern science was organized for projects of European expansion. Eurocentric interests, discursive resources, and ways of organizing the production of knowledge also constituted much of modern sciences in the first place.[26] Post-Kuhnian science studies provide many more examples of how religious, or national, or other local values, interests, discourses, and methods constituted modern scientific projects from their beginnings in early modern Europe.[27] In contrast to "intrusive politics," this kind of institutional politics does not force itself into purportedly pre-existing pure sciences; it has already constituted their natures and projects in the first place and on a continuing basis. State-of-the-art modern sciences always draw on local cultural resources in formulating their projects—for both better and worse.

Neutrality: From solution to problem. In this second case, where social relations help to *constitute* scientific projects as they together evolve, the neutrality ideal provides no resistance to the production of systematically distorted results of research, as we shall shortly see in more detail. But to put the matter this way is too mild a criticism of it. The neutrality ideal is not just useless in these circumstances; worse, it becomes part of the problem. Objectivism defends and legitimates the institutions and practices through which the distortions and their often exploitative consequences are generated. It certifies as value neutral, normal, natural, and therefore not political at all the policies and practices through which powerful groups can gain the information and explanations that they need to advance only their priorities—ones that usually conflict with others'.

Such information and explanations may well "work" in the sense of enabling prediction and control. However, this obvious fact does not provide evidence that the representation of nature's order in such a science is free of culturally local values and interests. A scientific account of one set of nature's regularities may at the same time obscure or draw attention away from other regularities and their causes, and from ones that would suggest other possibilities for human projects of organizing nature and social relations. One can get information about nature's order that makes possible building bigger bombs, accessing energy sources that are more toxic, or performing more expensive surgeries. Or one can get information that makes possible the equitable distribution of means to satisfy basic human needs for food, shelter, health, work, and just social relations of the world's economically and politically most vulnerable groups.

Moreover, the regularities of nature that make possible healing a body, charting the stars, or mining ores may be explained in ways permitting extensive (though not identical) prediction and control within radically different and even conflicting, culturally local, explanatory models. After all, Ptolemaic astronomy predicted perfectly well many of the movements in the heavens. Indeed, such observations have been retained in modern astronomy. Farmers in radically different cultures can predict equally well a large range of weather patterns and their effects, just as health care workers in these cultures can predict when illness will occur and how to cure it. The richness of cultural resources for grasping nature's regularities and their underlying causes is not exhausted by those favored in the modern North, as earlier chapters showed. Moreover, the kinds of explanations favored by modern science have not always been the most effective ones for all projects—for example, for maintaining environmental balance or preventing chronic bodily malfunctions. "It works" is no guarantee of cultural neutrality; nor do claims to cultural neutrality always accompany the best-working sciences and technologies.

The neutrality ideal functions more through what its normalizing procedures and concepts implicitly prioritize than through explicit directives. This kind of politics requires no informed consent by those who exercise it, but only that scientists be "company men" (and women), following the prevailing rules of scientific institutions and their intellectual traditions. This normalizing politics frequently defines the objections of its victims and any criticisms of its institutions, practices, or conceptual world as agitation by special interests that threatens to damage the neutrality of science and its promotion of social progress. This kind of politics makes visible just why it is that when sciences are already in the service of the mighty, scientific neutrality ensures that "might makes right."

It is many decades since it has been reasonable to think of modern natural and social sciences as small-scale, weak, guerrilla warfare projects for truth, struggling courageously against the evil empires of ignorance and superstition—Davids against the Goliaths. We need a concept of objectivity, and methods for maximizing it, that enable scientific projects to escape containment by the values, interests, discursive resources, and ways of organizing the production of knowledge characteristic of the kinds of powerful social tendencies identified by Levins and Lewontin. Objectivism's reliance on the neutrality ideal cannot do it. Such an analysis leads to one obvious possibility: to separate the goal of maximizing objectivity from the neutrality requirement in order to identify the knowledge-limiting values and interests that constitute projects in the first place. This possibility has been hinted at again and again in the literature without ever being formulated as a systematic program.

Before turning to examine such a program, let's examine in another way what is problematic about objectivism's only weak objectivity.

"Weak objectivity" cannot identify paradigms. From this perspective, the conventional notion of objectivity that links it to the neutrality ideal appears too weak to do what it sets out to do. That it is too weak is only

one way in which it is inadequate. But I use the term to acknowledge the usefulness of standards for objectivity-tied-to-neutrality in detecting the subset of distorting interests and values that do differ between individuals in the scientific community.

It is method that is supposed to "operationalize"neutrality and, thus, achieve objectivist standards, but method is conceptualized too narrowly to permit achievement of this goal. For one, method—in the sense in which students take methods courses or a research report describes its methods—is conceptualized as functioning only in the context of justification.[28] It comes into play only after a problem is identified as a scientific one, after central concepts, a hypothesis, and research design have already been selected. It is only after a research project is already *constituted* that methods of research, in the usual narrow sense of the term, start up. Moreover, the availability of a research technology that was itself selected in earlier contexts of discovery and found productive frequently helps select which scientific problems will be interesting to scientists and to funding agencies.

However, as critic after critic has pointed out, it is in the context of discovery that culture-wide assumptions shape the very statement and design of the research project, and therefore select the methods. Of course in the "mangle of practice"[29] during scientific research, hypotheses, nature, and research technologies are adjusted to each other such that a certain element of objectivity is produced without the promise of total neutrality. Nature constrains our beliefs without uniquely confirming them. The most science can hope for is results that are *consistent* with "how nature is," not ones that are uniquely *coherent* with it, as the objectivist goal intended.[30] Even the U.S. National Academy of Sciences—certainly not a den of wild-eyed radicals—now argues that the notion of research method should be enlarged beyond its familiar meaning of techniques to

> include the judgments scientists make about interpretation or reliability of data, . . . the decisions scientists make about which problems to pursue or when to conclude an investigation, . . . the ways scientists work with each other and exchange information.[31]

Thus, methods for maximizing objectivism have no way of detecting values, interests, discursive resources, and ways of organizing the production of knowledge that first constitute scientific problems, and then select central concepts, hypotheses to be tested, and research designs.

Let us approach the issue another way. One point of repeating observations, through experimental or other techniques, is so that variations in the results of observations can be scrutinized for the traces of social elements which would distort the image of nature produced by science. Any community that *is* a community, including the community of a laboratory or discipline as well as other kinds of cultural communities, shares values, interests, discursive traditions, and tendencies to relate to each other—to work together—in one way rather than another. But if all observers share a particular such cultural element, whether this arrives from the larger society or is developed in the group of legitimated observers, how is the repetition of observations by these like-minded people supposed to reveal it? It is not individual, personal, "subjective" error to which feminist and other social critics of science have drawn attention, but widely held androcentric, eurocentric, and bourgeois assumptions that have been virtually culture-wide across science. The assumptions of Ptolemaic astronomy, Aristotelian physics, or an organicist worldview were not fundamentally properties of individuals. Assumptions that women's biology, moral reason, intelligence, contributions to human evolution, or to history or present-day social relations are inferior to men's are not idiosyncratically held beliefs of individuals but widespread assumptions of entire cultures. The same is true for assumptions that non-European cultures are not capable of producing valuable sciences and technologies, that they are backward, and permeated by savagery and superstition, and that European cultures have largely developed with no significant contributions from any non-European cultures. Such assumptions have constituted whole fields of study, selecting their preoccupying problems, favored concepts, hypotheses, and research designs; these fields have in turn lent support to androcentric and eurocentric assumptions in other fields. The issue is not that individual men (and women) hold false beliefs, but that the conceptual structures of disciplines, their institutions, and related social policies make less than maximally objective assumptions.

In reflecting on how so much scientific racism and sexism could be produced by the most distinguished—and, in some cases, politically progressive—nineteenth-century scientists, historian of biology Stephen Jay Gould puts the point this way:

> I do not intend to contrast evil determinists who stray from the path of scientific objectivity with enlightened antideterminists who approach data with an open mind and therefore see truth. Rather I criticize the myth that science itself is an objective enterprise, done properly only when scientists can shuck the constraints of their culture and view the world as it really is. . . . Science, since people must do it, is a socially embedded activity. It progresses by hunch, vision, and intuition. Much of its change through time does not record a closer approach to absolute truth, but the alteration of cultural contexts that influence it so strongly.[32]

When a scientific community shares assumptions, there is little chance that more careful application of existing scientific methods will detect them.[33]

Moreover, Gould's reflection makes clear that not all cultural elements ("contexts") retard the growth of knowledge. Some advance it, he is saying: science has often progressed because of changes in its cultural contexts. So it is problematic that objectivism is supposed to eliminate *all* cultural elements. Weak objectivity is unable to discriminate between these aspects of culture that enlarge our understanding and those that limit it.

Are relativism and/or moral exhortations the only alternatives? The preceding section has identified some of the main features that make objectivism only "weak objectivity." When confronted with such issues, one apparent solution has been to turn to objectivism's Other, relativism,[34] sometimes with a kind of resignation that undermines both the critiques of objectivism and turn to relativism, at other times with the project of transforming relativism into a useful epistemological tool.[35] Excellent arguments both against objectivism and for relativism or subjectivism have been put forth by those who turn to this strategy. They are illuminating, and their arguments should not be undervalued. However, we need not examine them further in order to spot one great disadvantage that they have: relativism/subjectivism is the weak term in the objective/relative pair. Since, as the historians pointed out, appeals to these epistemological notions are primarily made as part of political struggles to claim this or that position for science in society, the weak term is unlikely to be attractive for these engagements. For those working in or close to the sciences, jurisprudence, or public policy, it will be very expensive to give up appealing to standards of objectivity and, instead, talk about the subjective or relative grounds for justifying beliefs, claims, and policies. Moreover, as we turn shortly to note, one cause of this weakness may well be that such alternatives to the neutrality of objectivism have been symbolized as feminine. Cultural definitions of manliness are at issue in turning away from objectivity-as-neutrality.

Yet another response has been to retain the neutrality criterion for maximizing objectivity, but to settle for moral exhortations that natural and social scientists should be more critical and that they should engage in dialogue with those protesting their exclusion from scientific authority. It is certainly better to have such moral gestures than not, since they can establish principles for belief and behavior that encourage movement toward blocking "might makes right." However, postcolonialism, feminism, and the other democracy-advancing social movements want and need more than this. It is unclear why women or the economically vulnerable citizens of developing societies should feel all that optimistic that the very groups whose interests and values were constituting distorting and oppressive research projects in the first place, sometimes unbeknownst to themselves, should want or know how to be more critical or engage in dialogue of the sort that would benefit the politically and economically most vulnerable.

So where might one find a *method* for maximizing objectivity that has the resources to detect, first, values, interests, discourses, and ways of organizing the production of knowledge that constitute scientific projects; second, ones that *do not* vary between legitimated members of research communities; and third, the difference between those cultural elements that enlarge and those that limit the adequacy of our descriptions and explanations of nature and social relations? This is where standpoint theory has provided useful resources that are not available— or, at least, not easily available—from other epistemologies.

Standpoint epistemology has been briefly described in earlier chapters and will be explored in greater detail in section 4 below and in the next chapter. First we turn to the special issues that arise for marginalized groups—issues beyond those discussed above—when notions such as objectivity, rationality, or good method are said to depend upon maximizing neutrality. We take women as an example of such a marginalized group.

3. Can women be objective?[36] We noted above that objectivity has been attributed to several distinct components of the research process. One of these was potential researchers: members of some races, classes, ethnicities, and genders have been claimed to be capable of greater objectivity than others. Not all groups have been thought capable of following the rules of scientific method equally rigorously. Manliness has been claimed to be more productive of objectivity than womanliness. Indeed, in so-called modernizing cultures, manliness has often been not just correlated with but defined in part through its capacity for the neutrality regarded as necessary for objectivity. Consequently, the results of men's research, and the processes used to arrive at them, have been able to bask in the beneficial status conferred on them precisely by their isolation from communities of women and their interests, discourses, and ways of organizing the production of knowledge. Critics point out that this is often the case regardless of the empirical or theoretical adequacy of such research results or the reasonableness of the methods used to arrive at them.[37]

The problem here may lie in the requirement that maximizing objectivity requires the maximization of neutrality. That is, insofar as objectivity is identified with neutrality, it appears impossible for women ever to be perceived as objective. This perception seems to persist regardless of how many women score high on mathematics tests or win Nobel Prizes. Briefly, objectivity has been thought to require neutrality; neutrality is coded masculine; and masculinity as individual identity and as symbolic meaning is culturally formed in opposition to the "feminine" and is continuously so maintained. Masculine is defined primarily by the absence of the traits attributed to the feminine. (Psychologists discuss this in terms of the aspects of the self that are projected onto the Other.) So how could women ever be perceived to be objective? This problem infects related concepts, also thought to require value-neutrality, such as "rational" and "scientific." It is worthwhile looking briefly at main themes in several well-known analyses supporting this argument.

Susan Bordo points to the ways in which the Cartesian age's "flight to objectivity" was a flight from the feminine—a defensive response to anxiety over the loss of the organic female universe of the Middle Ages and the Renaissance. "The Cartesian reconstruction of the world is a defiant gesture of independence from the female cosmos—a gesture that is at the same time compensation for a profound loss."[38] We should not, however, think of these meanings of masculinity and femininity only as mythologies, for

> the sexual division of labor within the family in the modern era has indeed fairly consistently reproduced these gender-related perspectives along sexual lines. . . . boys tend to grow up learning to experience the world like Cartesians, while girls do not, because of developmental asymmetries. . . . The historical identification of rationality and intelligence with the masculine modes of detachment, distance and clarity has disclosed its limitations, and it is necessary (and inevitable) that feminine modes should now appear as revealing more innovative, more humane, and more hopeful perspectives. (454, 456)

Genevieve Lloyd charted the trajectory of the association between the meanings of masculinity and rationality throughout the history of philosophy.[39] In a later paper she analyzes further the difficulty for women to be perceived to speak from the neutral position that is, paradoxically, also masculine.

> Seeing the maleness of reason is part of coming to understand how the symbolic structures work, realizing that there are speaking positions that, though supposedly gender-neutral, in fact depend on the male-female opposition.[40]

She draws attention to how examples of sexual symbolism can be only contingent on a philosophical argument, or somewhere between contingent on and constitutive of it, or sometimes fully constitutive of and deeply embedded in it. "Sexual symbolism operates in this embedded way in, for example, the conceptualization of reason as an attainment, as a transcending of the feminine" (82).

Evelyn Fox Keller notes that it is not only contemporary feminists who have pointed to the association of masculinity with objectivity. A century ago, George Simmel stated:

The requirements of . . . correctness in practical judgements and objectivity in theoretical knowledge . . . belong as it were in their form and their claims to humanity in general, but in their actual historical configuration they are masculine throughout. Supposing that we describe these things, viewed as absolute ideas, by the single word "objective," we then find that in the history of our race the equation objective = masculine is a valid one.[41]

When science is defined in terms of these linked meanings of objectivity and masculinity, not only is it difficult for women to speak within scientific discourses, but science itself is distorted.

The disengagement of our thinking about science from our notions of what is masculine could lead to a freeing of both from some of the rigidities to which they have been bound, with profound ramifications for both. Not only, for example, might science become more accessible to women, but, far more importantly, our very conception of "objective" could be freed from inappropriate constraints. (92–93)

Finally, Catharine MacKinnon shows how this linkage between objective, neutral, rational, and scientific on the one hand, and masculine on the other, deeply biases jurisprudence.

. . . the state will appear most relentless in imposing the male point of view when it comes closest to achieving its highest formal criterion of distanced aperspectivity. When it is most ruthlessly neutral, it will be most male; when it is most sex blind, it will be most blind to the sex of the standard being applied. When it most closely conforms to precedent, to "facts," to legislative intent, it will most closely enforce socially male norms and most thoroughly preclude questioning their content as having a point of view at all. Abstract rights will authorize the male experience of the world.[42]

Thus, "the feminist theory of knowledge is inextricable from the feminist critique of power because the male point of view forces itself upon the world as its way of apprehending it" (657).[43]

Yet one more observation of the links between masculinity and objectivity should be mentioned. Harry Brod has pointed out that ideals of masculine identity are highly dependent on public discourses; manliness, but not womanliness, is conceptualized in terms of activities and the meanings in the public realm. Consequently, disruption of public discourse on any topic regularly troubles men's senses of their masculine identity.[44] For example, when the U.S. economy weakens, misogynist discourses about the excessive costs of "social programs" for children, women, and the poor (all coded feminine) reach a higher volume. Or when the end of the Cold War results in decreased funding for scientific and technological research tied to military priorities, feminist science critics are blamed for a perceived national "flight from reason" and its purportedly accompanying decrease in science funding. So when skepticism about western standards of objectivity becomes centered in public discourses, the hidden object of skepticism is also the reasonableness of masculinity ideals.

Observations and arguments similar to many of these feminist ones appeared in the anti-racist and postcolonial literatures discussed in earlier chapters. Neither northern sciences' claims to value-neutrality nor their commitment to such a value can be value neutral, they pointed out: value-neutrality is not itself value neutral. An obvious consequence of this symbolic gendering, "racing," and "culturing" of objectivity and reason is to complicate attempts to resolve the objectivity question. We noted above some problems with attempts to substitute stronger forms of subjectivism or relativism for the flawed objectivist notion. Here are yet others: relativism and subjectivism are doomed from the start never to be able to achieve the kind of universal appeal of which their defenders dream since they carry ancient meanings of "not masculine" or "not European," and thus "not ideally human." On the other hand, how could feminism or the other liberatory discourses recover and transform a notion of objectivity that has been defined in terms of neutrality so that it will work for marginalized peoples, too, when it is constituted in the first place in opposition to the feminine, the "oriental," and so on? Choosing between weak objectivity and its neutrality ideal, on the one hand, or subjectivism/relativism, on the other, appears to be a lose/lose choice in these contexts.

Such reflections give added reason to try to delink maximizing objectivity from maximizing neutrality. By doing so, perhaps weak objectivity's hostility to the womanly and to Europe's Others can be mitigated by substituting a gender- and culture-ambiguous notion for the unrewarding choice between ones that have been coded fully masculine or fully feminine. Objectivity without neutrality can disrupt the gender and culture dimensions—and, more generally, the power relations—that persistently permeate the discourse of weak objectivity.[45]

4. *Strong objectivity.* How might we accomplish our tasks of systematically identifying interests, cultural discourses, and ways of organizing the production of knowledge that (co-)constitute scientific projects—that co-evolve with them and thus tend not to vary between legitimated observers—and of specifying the difference between those that enlarge and those that limit our knowledge?

First, we can reflect on the familiar observation that it will be easier to identify the contours of a given conceptual scheme or paradigm from "outside" than from within its categories, concepts, puzzles, and other preoccupations that usually fill up the entire horizon of our consciousness. We want to start off our thought from "elsewhere." Of course there is no absolute "elsewhere"; there is nowhere that is outside all culture. There are no vantage points anyone could find that are not themselves also discursively constructed within power relations: that was the delusion of weak objectivity and its neutrality ideal that must be abandoned. Nevertheless, we still have useful resources available, for we can start thinking from within another, different framework that enables us to look more objectively at our usually favored one. Where might we find one?

Here, we can reflect on a second familiar observation that ways of life or distinctive kinds of activities tend to give rise to distinctive ways of thinking and seeing the world. What we do both enables and limits what we can know. The sciences incorporate this insight in their understanding of the importance of having diverse methods as resources: each different kind of method enables the researcher to interact in different ways with the world, and so to have the chance to know its different features. What sciences enable us to know is what emerges from these different kinds of interactions with nature's order that the sciences ingeniously arrange. Earlier chapters explored how cultures' distinctive locations in heterogeneous nature, interests, discursive traditions, and ways of organizing the production of knowledge enable them to accumulate valuably different repositories of continuously revised empirical knowledge about nature and social relations. And earlier chapters explored how different locations in the political order—in gender or imperial relations, for example—generate different resources for the production of knowledge. Starting thought from such locations has enabled postcolonial and feminist science studies to detect some features of the prevailing conceptual frameworks of modern science that are otherwise difficult to see. Even the otherwise valuable post-Kuhnian science studies that failed to make use of postcolonial and feminist resources were unable to detect the eurocentric and androcentric assumptions that shaped their research. The immense value of postcolonial and feminist accounts has been provided by the way they take a perspective on scientific and technological change that lies further "out in culture" than the internalist perspective from which conventional intellectual histories of science had gazed at the history of science.

But what about our second task? Have we identified the difference between values and interests that advance and those that retard the growth of knowledge? Obviously, not every starting point for thought that lies outside a dominant conceptual framework is likely to enlarge our understandings. We can agree with the defenders of weak objectivity that at least *some* of the social elements that they think should be excluded from directing knowledge projects do indeed retard the growth of knowledge—"Think of Nazi science!" Others may be largely irrelevant to advancing the growth of knowledge; starting from the lives of, say, left-handed people—a devalued position in some cultures—may possibly reveal a little, but probably not much, about nature's order.

What we can do is look historically and cross culturally to see which kinds of shifts in the social climate, as Gould put the point, have enabled the detection of distorting, culture-wide social elements in patterns of knowledge. One important set of enabling social elements are those that resist "hiding" the most telling evidence against themselves. If women, the poor, and racial and ethnic "colonies" are kept illiterate, not permitted or encouraged to speak in public, and excluded from the design of the dominant institutions that shape their lives, they do not have the chance to develop and circulate their own politically and scientifically produced perspectives on nature and social relations. They do not have the chance to provide what could possibly be the most trenchant critiques of the dominant institutions and their discourses. Keeping the hypotheses that are most critical of the favored ones from ever getting to the starting line of a scientific research process obviously benefits the dominant groups. Clearly, the relations between the expansion of knowledge and the expansion of democratic social relations—always off-handedly claimed to be mutually supported by sciences' public discourses—need to be explored in greater detail. This democratic discourse of science conflicts with authoritarian practices such as those that have favored only weak standards for objectivity.

The next chapter explores further the research methods proposed by standpoint epistemologies for maximizing objectivity. Any suggestions for maximizing objectivity without neutrality could not provide a perfect solution to the "objectivity question" old or new, for there is no such solution that will do everything the liberatory social movements wanted of the older notion, and yet that will not seem strange and counterintuitive. We can recognize that what had seemed intuitively reasonable has become too problematic to retain; but acknowledging that fact does not lessen the strangeness of proposed alternatives. Our situation is similar in some ways to that of Galileo's culture. "Where did the crystal spheres go?" people asked. "How will God find us if we are not located in the center of his universe?"[46] Most likely, our questions will share the fate of these. Some will never be answered—at least not in the terms in which we pose them. Others will be answered within new theories of knowledge. The proposal here for standards for "strong objectivity" does solve some of the problems that plagued the older notions, even if it cannot answer all the questions the older epistemologies thought important to ask. Moreover, it captures intuitive insights of the post–World War II science studies projects.

Yet one more question may bother some readers. Why should we be concerned at all with trying to recover the notion of objectivity? Quite apart from the answers to this question proposed in this chapter and in previous ones—the value of elements in the European epistemological legacy, the reliance of the law, social policy, and so many other modern institutions and their self-images on such notions of objectivity as fairness, and so on—this line of thinking goes, the great power of "weak objectivity" was precisely its contradictoriness, its only selective functionality. By disabling the legitimacy of critically evaluating the broadest conceptual frameworks within which scientific issues are raised, it was possible for those benefitting most from dominant institutions and their discourses to practice "might makes right" while attributing the politicizing of knowledge projects only to the critics of current power-knowledge relations. From this perspective, producing a coherent notion of objectivity misses the point of such ideals, for it is precisely their contradictory nature that makes them useful. The notion of strong objectivity loses the benefits of such incoherence.

This is an illuminating insight. Indeed, all three schools of post–World War II science and technology studies shaping the arguments of this book have emerged precisely from the gaps in modern sciences' self-understanding. The workings and histories of modern sciences remain incomprehensible if one asks of them only the kinds of questions that internalist epistemologies have sanctioned. Those epistemologies delegitimated important kinds of questions that historians and sociologists think it appropriate to ask of every other social institution, its cultures and practices. The resulting mysteriousness of science is partly responsible for the quasi-religious status it achieves, as numerous commentators have recognized. So the "strong objectivity" project is part of this attempt to bring modern science within the domain of the kind of understanding we expect to be able to have of every other social institution. If the achievement of that goal makes rhetorical appeals to objectivity—"weak" or "strong"—less useful, so be it.

5. Truth. One more consequence of the demise of the neutrality ideal should be noted in closing this discussion. If more than one theory can reasonably be supported by any set of observations, and if every theory is reasonably open to more than one interpretation, is it still useful to invoke the notion of truth in conceptualizing an ideal relationship between our best knowledge claims and the natural and social relations that they are intended to describe, explain, or interpret? When the ideal results of research could be assumed to be socially neutral, truth or truth-approaching could appear to be a reasonable way to conceptualize the relationship. The best knowledge claims should be true of the world in the sense of reflecting without distortion the way the world is, of corresponding to a reality that is "out there" and unchanged by human study of it. Claims that satisfied the requirements of knowledge (that constitute "justified true belief") would bear a unique relationship to the world.

Yet science always promised something better than truth. It has always been understood that what makes a claim a scientific one, and not a matter of political dogma or religious faith, is that it is in perpetuity held open to revision on the basis of future, possibly disconfirming, observations and/or of revisions in the conceptual frameworks of the sciences. The abandonment in scientific circles of the concept of the crucial experiment in the late nineteenth century reflected the recognition that no empirical observations could prove a hypothesis true; (at most) they could only show it to be less false than its known competitors. Subsequently, Popper's deductivist logic and all the logical positivist focus on "verisimilitude" depended upon the assumption that truth could still function as a useful ideal as long as absolute falsifiability was possible. However, dreams of absolute falsifiability have proven unrealistic, also.[47]

Since facts—accepted empirical observations—are picked out as relevant ones by the theory they are supposed to be testing (including all of the background beliefs that support it) and by methods that are relatively inseparable from the theories that lead to their selection, facts can hardly stand as independent, value-, interest-, discourse-, and method-neutral tests of the empirical adequacy of the theory, as this chapter noted in opening. Observations are method- and theory-laden no less than theories are observation-laden, one could say. Historians have pointed out various kinds of situations where it was reasonable to retain a hypothesis in the face of apparent falsification. For example, as Kuhn's well-known account argued, it often makes good sense for young theories to be retained in the face of occasional or even frequent falsifying observations until they have developed more robust means of showing their empirical adequacy. Moreover, favored older theories turn out to be reasonably retained in spite of accumulating empirical evidence against them until there is an alternative available that appeals to the scientific community. In short, falsity is assigned to a theory not on the basis of some theory-neutral standard of the theory's inadequate fit with nature, but, instead, on the basis of a complex calculation by "the scientific community" as to when the potential benefits outweigh the potential losses in abandoning a hypothesis or theory.[48]

We can console ourselves by noting that abandonment of appeal to the truth of our knowledge claims does not commit us to relativism. After all, the procedures of the sciences (at their best) do generate claims that are validly and usefully regarded as "less false" in a limited (not absolute) but meaningful way: the hypothesis passing empirical and theoretical tests is less false than all (and only) the alternatives considered, though that judgment, too, must be held always only provisionally. Of course its status relative to hypotheses not yet tested is unknown.[49] And Gould pointed out the necessity of shifting social climates to enable vigorously critical alternative hypotheses to emerge.[50]

Thus, we could say that holding truth as an ideal for scientific claims as well as truth-maximizing procedures are not just unachievable but incoherent. What could it mean to imagine that all possible future observations and conceptual frameworks are available now for our critical scrutiny—that there is a finite supply of possible maps that cultures reasonably would find valuable to guide their interactions with the world, and that all such maps can be made not just consistent, but coherent with each other? The achievement of truth would mark not only the end of science, but also of history. Of course there are other criteria of truth; however, it is this correspondence criterion, paradoxically, that has most often been invoked in defense of modern sciences and their philosophies.

Truth claims are a way of closing down discussion, of ending critical dialogue, of invoking authoritarian standards. They deny the possibility of continuing processes of gaining knowledge in the future. We do not need them, for "less false" than the alternatives considered, tentatively held, will more accurately describe processes that have in fact resulted in the great achievements of modern sciences and those of other knowledge traditions.[51]

6. Conclusion. The notion of objectivity can be extracted from the neutrality requirement that has blocked its competence at advancing the growth of knowledge with respect to an important range of cases. When cultural elements constitute conceptual frameworks in the first place as sciences and their social orders co-evolve, one must start off thought from outside those frameworks in order even to identify them. Then the project still remains to figure out which of these cultural elements are at this particular historical moment advancing and which blocking the growth of knowledge. Adhering to the neutrality ideal in such cases insures only blind loyalty to unexamined cultural elements. It turns such elements into hidden evidence for the results of research so produced. We could reject the neutrality requirement as a nostalgic reminder of an innocent epistemological world that is gone from us forever.

9

Borderlands Epistemologies

1. Two problematic epistemological strategies. The new kinds of science and technology studies that emerged after World War II were created out of the systematic gaps in modern science's self-understanding, as the last chapter noted.[1] If modern sciences could have described and explained their own historical and social processes and the effects these have had on their understandings of nature's order in ways that made sense to historians and social scientists, there would have been no need for these science and technology studies. Of course no humans are ever able to understand fully "what we are doing," since we lack the historical long view, the awareness of larger economic, political, and social patterns, and an understanding of the causes of our collective fears, interests, preferences, and desires that subsequent histories, sociologies, political economies, and psychologies reveal.

To recap earlier discussions, in the early 1960s historian Thomas Kuhn had called for more of the kind of social history of modern science that could "display the historical integrity of . . . science in its own time." These histories were to explain scientific and technological change in ways that the prevailing intellectual histories could not. There was more history of science to be told than only the account of how succeeding generations of scientists used or revised their forerunners' ideas, which was the mainstay of conventional intellectual histories of science.[2] Of course these social histories would not turn out to be merely an additive project that left untouched conventional understandings of how the sciences have worked, for Kuhn's account revealed a different pattern to the growth of knowledge than the intellectual histories had detected. More than four decades of histories, sociologies, ethnographies, and philosophies of science have pursued this kind of project. Stimulated by the rising women's movements of the 1960s and 1970s as well as the post-Kuhnian paradigm, feminist scholars began to show the historical integrity of modern sciences with the gender relations of their eras. And earlier chapters in this study examined the postcolonial science and technology studies that have traced how modern sciences and technologies were fully integrated with economic, political, and social aspects of European expansion.

The older, "internalist" histories and philosophies of science that many of us learned were not just accidentally silent about such matters. Rather, they had denied the relevance and legitimacy of such accounts to understanding how science "really works." Such social accounts might explain how societies provided many of the resources scientific research requires, how they sometimes influenced the selection of which scientific projects were to be pursued, and how they applied the information science produced. And they could explain how politics sometimes managed to lead science down wrong paths, as with Lysenkoism and Nazi so-called science. But they could not explain science's successes, the internalist histories and epistemologies claimed, since these were the product not of such social factors, but of nature's order and the powerful features of scientific inquiry that lay entirely *inside* scientific processes. A distinctive scientific method of research, high standards of objectivity and of what can count as good reasoning, a critical attitude toward traditional belief, the distinctive metaphysics of nature that distinguished primary and secondary qualities, the use of mathematics to express nature's order, the particular way modern scientific communities have been organized—these and other features *internal* to science were especially suited to discovering nature's order. They were responsible for modern science's amazing successes, it was claimed. The post–World War II science and technology studies pointed out, however, that this internalist dogma, as some referred to it, left mysterious the answers to kinds of historical questions that are considered necessary to understand each and every other product of human activity.[3]

The internalist epistemology obscured why modern science emerged when and where it did, how it changed over time, and how its culture and practices co-evolved with those of other social institutions, such as the economy and state. Of course it was this epistemology of modern science—its method, standards of objectivity and rationality, the necessity of such features for social progress—that had long been used to justify the unique authority of the modern West in global political relations. Consequently, the discovery of these systematic and suspicious gaps were one source of rising global skepticism about the desirability and legitimacy of the authority of the West more generally. Widespread recognition of such failures has produced what is referred to as the epistemological crisis of the modern West.

What should the best theory of modern scientific knowledge look like now, more than three decades after the beginnings of the three kinds of post–World War II science and technology studies that this book has been examining? Two strategies for resolving the "crisis" that have been favored in post-Kuhnian science studies appear problematic from the perspective of postcolonial and feminist science and technology studies.[4] One project has been to try to patch up this conventional epistemology by responding to some criticisms of it and dismissing others without abandoning its fundamental internalist principle that the causes of the success of modern science are to be located entirely in nature's order plus science's internal processes. This project has seemed a good one to many philosophers of science; its assumptions also direct the practices of the natural and social sciences, as well as much public policy. These revisionists think that the prevailing epistemology of modern science should be retained in a modified form. It can simply be adjusted here and there in light of the representations of science that have emerged from the new histories, sociologies, and ethnographies of the sciences.

Another project, characteristic of many northern sociologists and ethnographers of sciences, is to agree with much more of the criticism of internalist epistemology—indeed, these theorists have themselves produced a great deal of it. But they presume that the only reasonable solution is to abandon epistemological projects completely and forswear the arrogance of presuming the political and intellectual appropriateness of the "policing of thought" that they think this requires.[5] They conceptualize all epistemological projects as necessarily internalist.[6] The only alternatives to such epistemologies are to be descriptive histories, sociologies, and ethnographies of science that disavow the normative stance taken by epistemologies. They try to substitute "sciences" of natural sciences, namely social sciences and more accurate historical accounts of natural sciences, for epistemologies of science. (One might ask if this is not itself an epistemological recommendation.)

It is noteworthy that these two approaches have succeeded in leading their defenders to think that a major target of their criticism should be each other.[7] While the discussions that have ensued are often illuminating, their preoccupations have blinded both to the possibility of more useful responses to the critiques of internalist epistemology than either patching up the latter or forswearing any epistemological recommendations at all.

While revised internalist epistemologies or the "abandonment" of epistemology have sometimes been the favored strategies in postcolonial and feminist science studies, a third approach has clearly emerged in both, which makes use of the resources of "borderlands" locations and states of mind. This is the approach of standpoint epistemologies, which has been referred to here and there in preceding chapters. This chapter pulls together into a more systematic sketch main standpoint themes. It concludes by distinguishing this epistemological approach from others with which it is sometimes conflated.[8]

2. What is a standpoint? Cultural and political epistemological resources. Standpoint epistemology has been most explicitly and extensively developed in recent decades within feminist theory. The concept of a standpoint arose from women's political struggles to see their concerns represented in public policy and in the natural and social science disciplines that have shaped such policy. These epistemologies propose that there are important resources for the production of knowledge to be found in starting off research projects from issues arising in women's lives rather than only from the dominant androcentric conceptual frameworks of the disciplines and the larger social order. Two kinds of "difference" provide independent arguments for abandoning the internalist epistemology of the modern West. One appeals to *politically* assigned locations in social hierarchies, such as those created by class, racism, imperialism, or sexism. The other appeals to *culturally* created locations, such as Chinese versus Puerto Rican, or Confucian versus Catholic. Though analytically distinguishable, in daily life the pervasiveness of political relations within and between cultures insures that different cultural resources almost never have equal political status.[9]

Intellectual and social histories of standpoint epistemologies. The intellectual history of feminist standpoint theory conventionally is traced to Hegel's reflections on what can be known about the master/slave relationship from the standpoint of the slave's life in contrast to the far more distorted understanding of it available from the perspective of the master's life. From the perspective of the master's activities, everything the slave does appears to be the consequence either of the master's will or of the slave's lazy and brutish nature. The slave does not appear fully human. However, from the standpoint of the slave's activities, one can see her smiling at the master when she in fact wishes to kill, playing lazy as the only form of resistance she can get away with, and scheming with the slave community to escape. The slave can be seen as fully human. Marx, Engels, and Lukacs subsequently developed this insight into the "standpoint of the proletariat," from which were produced theories of how class society operates.[10] In the 1970s, several feminist thinkers independently began reflecting on how the Marxian analysis could be transformed to explain how structural and symbolic gender relations had consequences for the production of knowledge. Subsequently, the resources of poststructuralism, anti-racism and postcolonialism became available so that some standpoint theorists could develop this "post-Marxist" epistemology in those directions.[11]

Standpoint arguments also appear in the knowledge projects of all of the new social movements, however.[12] A *social* history of standpoint theory would focus on the kinds of criticisms of prevailing institutions, their cultures, and practices that appear when formerly silenced peoples begin to gain public voice. On the one hand, these voices argue for applying the existing methods, rules, and procedures more fairly in order to eliminate what they think of as the biases in the prevailing views. However, they also frequently argue that the existing conceptual frameworks, methods, rules, and procedures for inquiry are themselves constituted only from the perspective of ruling-group interests. The standpoint of some particular marginalized group can point the way to less partial and distorted conceptual frameworks, methods, rules, and procedures of inquiry. What the standpoint of any particular group consists in must be determined by empirical observation and theoretical reflection. A standpoint is an objective position in social relations as articulated through one or another theory or discourse. Thus, such standpoint claims have recently been made by workers about the administrative/managerial class (no surprise, given the Marxian legacy of standpoint approaches), by critics of white supremacist practices in the United States, by gay and lesbian rights movements, as well as by postcolonial and feminist researchers, scholars, and activists.[13]

The conceptual practices of power.[14] Standpoint theory is not much concerned with the biases of individuals or of sub-groups within the dominant culture (one laboratory or research group versus another), which are the conventional focus of internalist epistemological thinking. Rather, its concern is with the assumptions generated by "ways of life" and apparent in discursive frameworks, conceptual schemes, and epistemes, within which entire dominant groups tend to think about nature and social relations, and to use such frameworks to structure social relations for the rest of us, too. That we all have to live within the social worlds created by elites with distorted understandings of how society works makes it extra difficult to achieve a standpoint. A standpoint is not the same as a viewpoint or perspective, for it requires both science and political struggle, as Nancy Hartsock puts the point, to see beneath the surfaces of social life to the "realities" that structure it.[15]

We can pick out several major themes in standpoint approaches. First, the starting point of standpoint theory, and its claim that is most often misread, is that in societies stratified by race, ethnicity, class, gender, sexuality, or some other such politics shaping the very structure and meanings of social relations, the *activities* or lives ("labor" in the Marxian account) of those at the top both organize and set limits on what persons who perform such activities can understand about themselves and the world around them. "There are some perspectives on society from which, however well-intentioned one may be, the real relations of humans with each other and with the natural world are not visible."[16] In contrast, the activities of those who are exploited by such social hierarchies can provide starting points for thought—for everyone's research and scholarship—from which otherwise-obscured relations that people have with each other and with the natural world can become visible.

Chapter 4 outlined how different cultures are led to ask different questions about nature and social relations because of their distinctive locations in the natural world (in deserts, on waterways, in the Arctic, or on the equator), their distinctive cultural interests even in "the same" environment, their culturally local discursive legacies (the metaphors, models, narratives, and the like through which they have defined themselves as a culture and come to see the world around them), and their distinctive ways of organizing the work of producing knowledge. Chinese and Puerto Rican patterns of knowledge and ignorance will differ because of such cultural differences. However, power differences within or between cultures will also create different opportunities for systematic knowledge and systematic ignorance. The experience and lives of marginalized peoples, as they understand them, provide distinctive *problems to be explained* or research agendas that are not visible or not compelling to the dominant groups. Marginalized experiences and lives have been devalued or ignored as a source of important questions about nature and social relations, especially objectivity-maximizing ones, as the last chapter explored. It is valuable new questions that thinking from the perspective of such lives can generate.

However, the answers to such questions are never completely to be found in those experiences or lives. For the answers, one must examine critically the dominant conceptual frameworks that reflect disproportionately the interests of dominant groups. It is dominant groups who, in making what appear to them to be perfectly reasonable policies, shape marginal lives in ways not always visible within those lives. For example, women, too, have tended to see their household labor as not really work, and to be unable to see what is now referred to as sexual harassment as a violation of their civil rights. "I never called it rape," is the title of a well-known study of marital and date rape. And why would anyone call such events rape when the law held that a wife gave up the right to say "no" upon entering marriage, and when courtship rituals dictated that women's protesting "no" always meant "yes"? Thus, standpoint theories argue that it is certainly the case that each group's social situation enables and sets limits on what it can know. However the critically unexamined dominant ones tend to be more limiting than others in this respect. What makes these social locations more limiting is their inability to generate—indeed, their interests in avoiding, devaluing, silencing—the most critical questions about the dominant conceptual frameworks. Marginalized groups have interests in asking such questions, and dominant groups have interests in not hearing them.

Of course this does not mean that all women will be able to ask the most critical questions of androcentric frameworks. Obviously such feminist questions only occasionally arise in history. Even when they have come up with such questions, most women have usually correctly assessed the immense personal cost of challenging the male-suprema-cist norms in different cultures. Nor does it mean that no men can ever ask them; there are plenty of examples of men doing so in the history of feminist political activism, research, and scholarship. Many men have found it more personally costly to continue blindly adhering to male supremacist norms than to speak out and act against them, and they have gone on to construct feminisms of their own.[17] Standpoint theory is only pointing to how people tend to perceive their own best interests in predictable ways, though they can always find reason to pursue other interests that make exceptions to such predictions.

It is this sense in which Dorothy Smith argues that women's ex-perience is the "grounds" of feminist knowledge and that such knowl-edge should change the dominant conceptual frameworks of sociology. Women's lives (our many different lives and different experiences) can provide the starting point for asking new, critical questions about not only those lives, but also about men's lives and the social institutions designed primarily by men to serve "humanity." Most importantly, Smith argues, a sociology that is to be *for* women, rather than for the dominant social institutions and their beneficiaries must ask new ques-tions about the causal relations between women's lives, on the one hand, and men's lives and public institutions, on the other hand.

For example, she points out that if we start thinking from women's lives, we (anyone) can see that women are assigned the work that men do not want to do for themselves, especially the care of everyone's bodies— the bodies of men, of babies and children, of old people, of the sick, and of their own bodies. And they are assigned responsibility for the local places where those bodies exist as they clean and maintain their own and others' houses and workplaces. Some women are assigned more of this work than others—much, much more; but even wealthy and aristo-cratic women with plenty of servants are left significantly responsible for such work in ways their brothers are not. And men in marginalized groups often perform certain kinds of such work in restaurants, hospi-tals, and janitorial jobs. This kind of work, she shows, frees men in the ruling groups to immerse themselves in the world of abstract concepts. The more successful women are at this concrete work, the more invisible it becomes to men as distinctively social labor. Caring for bodies and for the places in which bodies exist disappears into nature. Consider, for example, sociobiological claims about the naturalness of altruistic be-havior and domestic work for females and the unnaturalness of either for males. Or consider the systematic reticence of many prefeminist Marxists to analyze who does what "labor" in everyday sexual, emo-tional, and domestic work and to integrate such analyses into their accounts of "working class labor." Smith argues that we should not be surprised that administrative/managerial men have trouble seeing women's activities as part of distinctively human culture and history once we notice how invisible the social character of this work is from the perspective of such men's activities.

She points out that if we start from women's lives, we can generate questions about why primarily women are assigned such activities and what the consequences are for the economy, the state, the family, the educational system, and other social institutions of assigning body and emotional work to one group and head work to another.[18] Such questions lead to less partial and distorted understandings of women's worlds, men's worlds, and the causal relations between them than do questions originating only in that part of human activity that men in the dominant groups reserve for themselves—the abstract mental work of management and administration.[19]

Similar accounts of the tendency of the eurocentric, colonial, or imperial mentality to conceptualize "natives" as part of nature, of their labor as not really social labor, not really part of human history, their land as empty or wasteland, are common.[20] Recent changes in the international division of labor increase the apparent reasonableness of this kind of distorted conceptual framework, for much of the most degraded manufacturing and agricultural labor that formerly was conducted within national borders in the North, where it was at least somewhat visible to northerners, now has now been relocated to the free trade zones or other parts of the developing world, where it is much less obvious to northern eyes.

Thus, standpoint epistemology sets the relationship between knowledge and politics at the center of its account in the sense that it tries to explain the effects that different kinds of political arrangements have on the production of knowledge. Of course, the older empiricist theories of knowledge were also concerned with the effects politics have on the production of knowledge, but prefeminist empiricism conceptualizes politics as entirely a threat to the purity of scientific knowledge.[21] Empiricism tries to purify science of all such bad politics by adherence to what it takes to be rigorous methods for the testing of hypotheses. For reasons explained in the last chapter, this is far too weak a strategy to maximize the objectivity of the results of research that empiricists desire. Thought that begins from the lives of the marginalized has no chance to get its critical questions voiced or heard within such an empiricist conception of the way to produce knowledge, nor can the positive value of such "political" questions be detected within empiricist frameworks. Prefeminist empiricists can only perceive such questions as the intrusion of politics into science, and therefore as deteriorating the objectivity of the results of research, as was discussed in the last chapter.

Thus, the standpoint claims that all knowledge attempts are socially situated, and that some of these objective social locations are better than others as starting points for knowledge projects, challenges some of the most fundamental assumptions of the scientific worldview and the western thought that take science as their model of how to produce knowledge. It sets out guidelines for a "logic of discovery" intended to maximize the objectivity of the results of research, and thereby to produce knowledge that can be for marginalized people and for those who would know what they can know, rather than for the use only of dominant groups in their projects of administering and managing marginalized lives.

How can the lives of "the weak" provide resources for the growth of knowledge in the natural sciences? Posing the question this way draws attention to the political differences—rather than merely cultural ones—that can be used to advance the growth of knowledge. Elsewhere I identified a number of aspects of hierarchical social relations that can be expected to produce distinctive scientific and epistemic resources for "the weak." These knowledge opportunities are already widely recognized in the literatures of different social science and humanities disciplines.[22] That is, this standpoint claim that scientific advantage accrues to those who start asking questions about nature and social relations from the social position of "the weak" turns out not to be so radical or new after all.

How does this work for the natural sciences? Most conventional histories, sociologies, and philosophies of science assumed that science's social relations are constituted fundamentally by public, official, visible, and dramatic role players and situations—scientists and their critics who were recognized as such in their own day, for example. "Yet unofficial, supportive, less dramatic, private and invisible spheres of social life and organization may be equally important," as the editors of one of the early influential feminist critiques of sociological theory and practice pointed out.[23] Of course, it is those who are public, official, visible, and dramatic role players, or who see their interests as well represented in those kinds of situations, who would make such limited and distorting assumptions about the constitution of a culture such as scientific culture. "Our activities are the only ones that count as social," they seem to say. So it was by starting off analyses from the unofficial, supportive, less dramatic, private, and invisible spheres of social life "outside science," ones that support and sustain public, official, visible, and dramatic *scientific* role players and organizations, that it has been possible to produce more accurate and comprehensive accounts of the historical integrity of sciences with their cultures. Feminist and postcolonial critiques of the conceptual frameworks of each discipline contributing to the social studies of science—history, economics, political philosophy, anthropology, sociology, psychology—draw our attention to unacknowledged aspects of the culture and practices of both modern sciences and the scientific and technological traditions of other cultures, as we saw in earlier chapters.

This counterdominant perspective has also been analyzed in history and sociology as the standpoint of everyday life, as indicated earlier. One can learn much by starting off thought from the lives of those who perform the daily routines necessary for everyone's bodily and social survival, thereby relieving members of dominant groups of the need to take responsibility for the care of their own bodies and daily social needs. Much of the discussion of the role of local knowledge systems focusses on just such epistemic advantages. Insofar as different groups are assigned different daily activities, they will tend to know different things about natural and social worlds.

Another way this "power of the weak" has been discussed is in terms of the advantages of the stranger or outsider. The stranger brings to research just the combination of nearness and remoteness, concern and indifference, that are central to maximizing objectivity.[24] Women, racial/ethnic minorities, the victims of imperialism and colonialism, and the poor are in some respects functionally "strangers" to the dominant cultures and practices that structure their lives—including such scientific and technological cultures and practices. Their needs and desires are not the ones that have found expression in the design and functioning of the dominant institutions. It is not their needs and desires that have designed the forms of general and professional science education to which they are subjected, the priorities of funding for scientific and technological research, or the design of scientific and technological changes. And yet these groups are not completely outside the dominant institutions—they are no longer off in Africa or barefoot and pregnant in the kitchen. They are instead on the margins, the periphery; they are "outsiders within" or on the "borderlands," in two influential standpoint phrases.[25]

Again, the standpoint from such marginalized lives on whether and how scientific and technological changes work can lead to a more objective account than can analyses restricted to what looks reasonable from the perspective of the groups who most benefit from scientific and technological change. "The winner names the age" the historians say, acknowledging that the winners' name for the age may not be the most accurate one from more objective standpoints. Starting from the "losers'" lives can systematically expand our knowledge. This is one way of talking about how people marginalized, dominated, oppressed, or otherwise disadvantaged by a dominant culture have fewer interests in ignorance about how such a culture and its practices actually work than do those that benefit from it. Anyone who starts out thinking about science funding, or environmental destruction, or medical research from the perspective of the lives of those who bear a disproportionate share of the costs of these activities can learn to "follow the interests" of the latter to arrive at less partial and distorted accounts of science and technology institutions and practices.

Historians have come to understand the importance of "underclass history" as they tell labor history from the perspective of workers' interests, the Civil War from the perspective of the interests of foot soldiers and of African Americans on both sides, the history of feudal societies from the perspective of the interests of peasants, and European expansion from the perspective of the interests of the cultures into which Europe expanded. In earlier chapters we have seen more accurate and comprehensive accounts of the history of European and non-European scientific and technological traditions told from the perspective of the interests of the latter.

The older Marxian accounts argued that certain social formations only became easily visible at certain historical moments. They were interested in explaining why it was only in the 1840s and thereafter that the class system could begin to be perceived as a social force that changed independently of other social and natural formations—as the kind of relationship between proletariat and bourgeoisie that caused other social features rather than being only an effect of them, as earlier accounts held. Feminist theorists have described the emergence into visibility only after World War II of what has been called the "gender system"—a system that is not entirely an effect of biology, of class relations, or of some other social arrangements.[26] Of course class and gender relations are far older than the 1840s and 1950s, respectively; but their relative independence from other social formations only becomes visible at these points. Similarly, we could say that the causal role of colonial/imperial relations in shaping the growth of modern scientific and technological change only became visible with the end of formal political rule of their colonies by European nations—though there were intimations of these analyses-to-come several decades earlier (speech that in its very possibility heralded the looming end of such political rule).[27] Thus, epistemic advantage with respect to any particular social formation of sciences and technologies can wax and wane at different historical moments. The ability to identify and think from those sites— to identify the contradictions within the dominant ways of organizing social life—is to enhance one's chances for more accurate and comprehensive accounts of nature and social relations.

Finally, a whole range of interpretive strategies in literary, cultural, and historical studies draws our attention to alternative readings not only of conventional texts—of spoken or written words—but also of cultural formations. They show how to analyze just who is the implied "subject" of scientific claims, the cultural assumptions of the metaphors, models, and narratives in scientific accounts, and how the meanings of scientific claims are created at the conjunction of scientific culture and practices, a particular historical moment, and a particular group of people interacting with sciences and technologies.[28] Starting from marginalized lives makes it easier to see the discursive formations that construct and continuously relegitimate dominant conceptual frameworks in the sciences and the larger societies that evolve together.

To summarize my point here, existing analyses in the social sciences and humanities offer many resources for understanding how lives that have been marginalized by the dominant conceptual frameworks can turn out to provide otherwise unavailable resources for better describing and explaining the natural and social relations that are the object of science and science studies accounts. This central claim of standpoint epistemologies turns out to be not quite as surprising as it might initially appear. However, from the perspective of the pervasive general eurocentric and androcentric mind-set of modern scientific rationality, such an epistemology remains difficult to understand and appreciate. Our choices of epistemologies have also been restricted to the ones that look most reasonable to dominant groups through the same kinds of processes responsible for distorted conceptual frameworks within natural and social science research fields.

We can understand a bit more about what a standpoint is by separating these epistemologies from some common misreadings of and apprehensions about them.

3. What standpoint epistemologies are not. For those who still hold that maximizing objectivity requires maximizing neutrality (the problematic "weak objectivity" discussed in the last chapter), standpoint epistemologies will appear relativist. From such a perspective they appear as a kind of special pleading or unreasonably claimed privileged positionality. On such a reading, empiricism is politics-free, and standpoint theory is asserting epistemological/scientific privilege for one group at the expense of the equally valuable and/or equally distorted perceptions of other groups. All groups are "biased," they are willing to admit, so standpoint approaches are simply claiming privilege for one kind of such bias. They are substituting one politics for another, for all political positions—the master's and the slave's, that of the rich and of the poor, of the colonizer and the colonized, of the rapist and his victim—are equally valuable and/or equally biased in the production of knowledge.

This interpretation of difference as mere diversity is a serious misunderstanding of social relations, as well as of political standpoint claims about the effects of power on knowledge claims. It reduces power relations to mere cultural differences. Standpoint theory leads us to turn such a way of posing the issue into a topic for historical analysis: "what forms of social relations make this conceptual framework—the 'view

from nowhere' versus 'special pleading'—so useful, and for what purposes?" Let us look at some of these common misunderstandings of
standpoint theory.

Not only about marginal lives. First, standpoint theory is not only about
how to get a more accurate understanding of marginal lives—the
peoples European expansion encountered, or European and North
American women's lives, for example. Instead, research is to *start off*
from such locations in order to explain the relationship between those
lives and the rest of social relations, including human interactions with
nature. The standpoint of women, as we saw Dorothy Smith put the
point earlier, enables us to understand women's lives, the social institutions from the design of which women are excluded, and the relations
between the two. To do this we need concepts and hypotheses arising
from women's lives (though not necessarily the ones women themselves
use, which often are part of the dominant institutions' ideologies),
rather than only ones arising from the lives of those assigned adminis-
trative/managerial work, a group that includes sociologists (and philosophers). The point is to produce systematic causal accounts of how
the natural and social orders are organized such that the everyday lives
of women and men, Europeans and those they encounter, end up in
the forms that they do.

"Grounded," but unconventionally so. The phrases "peasant experiences" or "women's experiences" can be read in an empiricist way such
that these experiences are assumed to be constituted prior to the social.
Major strains of standpoint theory challenge this kind of reading.[29] For a
researcher to start from marginalized lives is not necessarily to take
one's research problems in the terms in which marginalized people
perceive or articulate their problems—and this is as true for researchers
who come from these groups as it is for those who do not. The dominant
discourses, their institutions, practices, favored conceptual frameworks,
and languages, restrict what everyone is permitted to see and shape
everyone's consciousness. Fortunately, they are not perfect at these
projects, for subjugated discourses always also exist; power always also
produces in its subjects visions, dreams, plans for its end. Women, like
men, have had to learn to think of their domestic work not just as a
"labor of love," but as a contribution to the local and national economy.
Many citizens of so-called developing countries have had to reassess just
who is benefitting from the "progress" that the transfer to their cultures of modern sciences and technologies is supposed to be bringing.
Women, too, have held distorted beliefs about our bodies, our minds,
nature, and society, and numerous men have made important contributions to feminist analyses—John Stuart Mill, Marx, Engels, and many
contemporary scholars in history, sociology, economics, philosophy,
jurisprudence, literary and art criticism, and other fields. It is obvious

that "peasant experience" or "women's experience" does not automatically generate counterhegemonic analyses, since the former often exists but only occasionally does the latter emerge. Standpoint theorists are not making the absurd claim that the new postcolonial and feminist analyses simply flow naturally from these groups' experiences.

Postcolonial and feminist analyses are not culturally neutral elaborations of people's social experiences, or what members of marginalized groups say about their lives; they are theoretical reflections on them. Marginalized experiences, and what marginalized peoples say, are crucial guides to the new questions that can be asked about nature, sciences, and social relations. Such questions arise out of the gap between marginalized interests and consciousness, on the one hand, and the way the dominant conceptual schemes organize social relations, including those of scientific and technological change. Moreover, the answers to such research questions cannot be found simply by examining more carefully marginalized lives, since marginalized interests and experiences are shaped by national and international policies and practices that are formulated and enacted far away from marginalized peoples' daily lives. The everyday lives of marginalized peoples are shaped by National Science Foundation research priorities, by international environmental agreements, by policies adopted in the International Monetary Fund or World Bank, by military policies made on the other side of the world, and so on. Standpoint theory is not calling for phenomenologies or merely rational interpretations of marginalized worlds. Nor is it arguing that only members of marginalized groups can generate knowledge that is useful to such groups—that is *for* them. Standpoint epistemology is not an "identity politics" project for knowledge production—unless "identity" is taken as one's commitment to who one wants to be rather than only to where one has come from.

Men, too, can learn to start their thought from women's lives, and northern peoples from southern lives, as many have done. Misunderstandings come about because objectivism insists that the only alternative to its "view from nowhere" is special interest biases and ethnoknowledges that can be understood only within a relativist epistemology. In contrast, standpoint epistemologies propose that institutionalized power imbalances give the act of starting off from marginalized lives a critical edge for formulating new questions that can expand everyone's knowledge about institutionalized power and its effects. Postcolonial and feminist science and technology studies have undertaken just such projects, as earlier chapters explored.

Much of the debate over just what "grounds" standpoint accounts is a consequence of the different ways feminist (and postcolonial) analyses have theorized what was problematic about internalist epistemologies. Where the internalist epistemologies insisted on the scientific priority, the greater objectivity, of outsiders' descriptions over "the natives," the new social movements have insisted that such descriptions greatly lacked objectivity, and that marginalized groups should get to express their concerns in their own terms. Their "experiences" were at least as good as the experiences of their "masters" in providing objective descriptions and explanations of the social relations between them. There was no innocent, disembodied view from nowhere possible with respect to gender, class, or race relations in which everyone was implicated, whether or not they chose to recognize such loss of innocence. The site from which more objective analyses were to emerge was not individuals' consciousness, but collective histories. One's position in such histories was crucial. The most objective knowledge claims were neither to be determined by one's "material" location, nor to have any relation at all to such a location—to be merely "ideal"—and so forth. My point here is that as different aspects of the internalist epistemology were problematized, the standpoint accounts tended to shift across claims about women's experiences, "lives" or historical conditions, embodiment, sites or positions in social relations, feminist discourses, and various other "grounds" for knowledge forbidden in the internalist epistemologies. The suggestion here is that it is more important to understand how the pattern of these claims was created by what they were opposed to and how they were devalued in the mainstream epistemologies than to try to settle on one or another as the really only defensible one.

No essential marginalized lives. Next, standpoint theory is not arguing that there is some kind of essential, universally adequate model of the marginalized life from which research should start off. In any particular research situation, one is to start off thought from some particular set of lives that have been exploited by those who favor a dominant conceptual frameworks. What can we learn about that framework by starting off from the activities that are the preconditions for dominant group activity, but unacknowledged as such in the dominant conceptual schemes? For example, what can we learn about biological models of the human body by starting off from the diverse activities of women, poor people, and racially marginalized groups whose bodily activities in households, offices, restaurants, and other places where biologists sleep, work, and live are the precondition for biologists' having the leisure to pursue their studies? These are socially diverse groups of people whose different kinds of work make possible the pursuit of biological research. "Racially marginalized," "poor," and "women" are not homogenous categories; they include groups whose activities are differently shaped by their class, race, gender, ethnicity, historical period, and cultural milieu. Any presumption of uniform experiences and activities would distort the accounts that ensued. Though the conventional way of thinking about power relations tends to enshrine an oppositional, two-party relationship between a homogenized "us" and a homogenized "them," power functions in far more complex ways.

Consciousness not determined by social location. According to standpoint theorists, we each have a specific location, albeit often a complex one, in such a social matrix; but that location does not determine one's consciousness. The availability of competing discourses enables some men, for example, to think and act in feminist ways. Yet they still obviously remain men, who are thereby in determinate relations to women and men in every class and race. They can work to eliminate male supremacy, but no matter what they do, they will still be treated with the privilege (or suspicion!) accorded to men by students, sales people, coworkers, family members, and others. The same can be said of anti-eurocentric people of European descent. Parallel accounts can be given about other groups of marginalized peoples, of course.[30] The point of standpoint theory is to help move people toward liberatory standpoints, whether one is in a marginalized or dominant social location. It is an achievement, not a "natural property," of women to develop a feminist standpoint, or a standpoint of women, no less than it is for a man to do so.

An epistemology, a philosophy of science, a sociology of knowledge, and a method for doing research. Several disciplines have competed to disown (and in one case not only to claim, but to monopolize) standpoint theory. Some philosophers and natural and social scientists see it as only political advocacy. Other philosophers claim it is only a sociology; some political scientists and sociologists reject it as only an epistemology, though one sociologist has insisted that it is not an epistemology at all, but a method for doing sociology.[31]

Reflection on such rejections of standpoint theory can be illuminating, for they reveal how severely this theory diverges from the standard disciplinary models that conceptualize representations of knowledge seeking. It is more useful, I suggest, to see it as all of these projects: a philosophy of knowledge, a philosophy of science, a sociology of knowledge, a moral/political advocacy of the expansion of democratic rights to participate in making the social decisions that will affect one's life, and a proposed research method for the natural and social sciences. Each such project must always make assumptions about the others. For example, every philosophy of science must make epistemological assumptions about the nature of and conditions for knowledge in general, historical ones about which procedures for producing knowledge have been most successful in the past, sociological ones about how communities that have produced the best knowledge claims in the past have been organized, and moral/political ones about which humans should have full democratic rights. In periods of what we could refer to as "normal philosophy," these background assumptions can safely be left unexamined; but when skepticism arises about the adequacy of fundamental assumptions in any one of these areas, the others all present themselves as candidates for reexamination. Our beliefs face the tribunal of experience as a network, and none are immune from possible revision when a misfit between belief and observation arises, as philosopher W. V. O. Quine put the point.[32] Postcolonial, feminist, and post-Kuhnian social studies of science and technology have been raising challenges to conventional conceptual frameworks that have led to reexamination of empiricist assumptions about the organization of scientific communities, ideals of the knower, the known, and how knowledge should be produced, rational reconstructions of the growth of scientific knowledge, and scientific method in the sense of "how to do good research." Standpoint theory's claims have effects on and must draw resources from all of these fields.

Not damagingly relativist. As mentioned above, standpoint approaches are not relativist or pluralist in the usual sense of these terms. In everyday talk, "standpoint" is used interchangeably with "view," "perspective," and other such locational terms that are relativist not only in that they are socially located, but also in that all have equal authority; none is inherently more advantaged or privileged than any other. However, in the originating analyses on which standpoint theorists reflected, starting off thought from the master's life was not just as good as starting off from the slave's life to understand the master/slave relationship. Nor was the view from bourgeois lives supposed to be just as good as the view from workers' lives to explain how capitalist economic relations worked. In these cases, the exploited social position offered the possibility of a critical perspective on the dominant institutional and conceptual systems. Thanks to African American history and labor history we have come to understand systems of slavery and of class societies in ways that were not visible from the lives of those benefitting from such systems.

Claims can be sociologically or historically "relativist" in the sense of locating a distinctive pattern of thought in its historical and social context: different cultures (classes, genders, historical epochs) tend to favor different patterns of thought. But that still leaves us with the possibility of adopting a position of cognitive or epistemological relativism, or not; it does not force us to a relativist position. The relativist would say that each of these historically specific patterns of thought—these local knowledge systems, as earlier chapters referred to them—is equally good, or true, or valid. For what, one can ask? For getting to the moon? For sustaining fragile environments, or democratic social relations? The absolutist would say that some of these patterns can be known to be universally good. Again one wants to ask, for what? Is there a single pattern of knowledge equally good at all challenges any human culture might set for itself? These two conventional positions do not seem to be addressing the kinds of issues that have been the concern of postcolonial, feminist, and post-Kuhnian science and technology studies. The kind of epistemology explored and recommended in this book argues that such different local knowledge systems each have their own distinctive resources for and limitations on understanding ourselves and the natural and social worlds around us. The practical challenge raised by post–World War II science and technology studies is to understand which are the resources and which the limitations for any given knowledge system, and which systems are best for which knowledge production projects.

4. *Conclusion.* Adopting a standpoint epistemology or method of research (in the broad sense of "method") may often be the only way to detect the broad, cultural, or civilizational conceptual frameworks limiting certain kinds of knowledge-seeking projects. How else could we who live in societies structured by eurocentrism and androcentrism, for example, come to detect the more pervasive forms of such elements shaping our thinking and actions? Standpoint approaches can show us how to detect values and interests that constitute scientific projects, ones that do not vary between legitimated observers, and the difference between those values and interests that enlarge and those that limit our descriptions, explanations, and understandings of nature and social relations. Standpoint approaches provide a map, a method, for maximizing a "strong objectivity" in the natural and social sciences. They provide more objective ways of explaining the limitations of standard accounts of nature and social relations, and the surprising strengths of the post-Kuhnian, postcolonial, and feminist studies of science and technology that have emerged since World War II.

However, they may not always be the best way to articulate why a particular knowledge seeking strategy is preferable. After all, the point of an epistemology is to explain the reasonableness of a particular understanding of knowledge seeking. Apart from other considerations, an epistemology should be plausible. And the kind of account of how political relations shape the production of knowledge may well be incomprehensible to many who still cling to the older empiricist and other internalist epistemologies. Much that standpoint epistemologies explain can be explained by conventional empiricist appeals to "good method," "rigorous observation," "eliminating distorting social biases," and the like. Much will not be explained through such a conceptual framework, as indicated throughout this chapter. My point here is that the preference for one epistemology over another can reasonably be as strategic as the preference for one scientific theory over another: it provides the kind of map we need to get us where we want to go. Of course we then must justify why it is *there* that we want to go. Border-lands have emerged as expanding and crowded territories of contempo-rary social life. Standpoint epistemologies articulate how important forms of knowledge can be produced from such "territories."

What has happened to the universality ideal in these accounts? If we have only local knowledge systems, how can different cultures commu-nicate with each other and work together in useful ways? And how does standpoint theory account for itself? Standpoint theories, too, must be located in their local histories and social relations. How do they deal with the issue of reflexivity that has troubled late-twentieth-century studies of the production of knowledge? We turn to these issues next.

10

Dysfunctional Universality Claims?
Scientific, Epistemological, and Political Issues

1. Universality and modernity. The ability to produce a uniquely universal science is commonly thought to be a distinctive mark of modernity. Such a science should have uniquely valid standards of rationality, objectivity, method, and what counts as nature's order and as knowers. Such purported features of modern science have been called on to legitimate as uniquely socially progressive European models of government, law, education, social policy, and even ethics. From this perspective, modern Europe and its diasporas in the Americas, Australia, and elsewhere provide a uniquely desirable model of human achievements, social relations, and standards of living.

Other cultures' accounts of themselves and the world around them, like those of premodern Europe, cannot really be considered in the same category as modern sciences, this account holds. They should not even be called science traditions since there is one and only one possible true account of nature, and that is the one modern science has been struggling to piece together.[1] As noted in earlier chapters, such other accounts are to be regarded only as "ethnosciences," "folk thought," "precursors of modern science," belief systems "with scientific elements," or—worse—savage and barbaric thought, witchcraft, superstitions, magic, and products of pre-logical mentalities.

Moreover, models of objectivity, rationality, scientific method, and their ability to advance modernity have also invariably been defined in terms of idealizations not only of Europeans, but also of masculinity, as the feminist critiques reviewed in earlier chapters have shown. The desirable virility of European civilization is often signified by the progressiveness of its modern sciences, and the desirability of dominant models of manliness by their links to modernity, its rationality and social progressiveness. The thinking and behavior of women of European descent, too, are to be assigned to the premodern, according to this kind of conventional thinking.[2] Thus, the meanings of claims to modern science's universality are constructed in opposition to meanings of womanliness and cultural otherness.

The power of these meanings is indicated by their immunity to empirical evidence to the contrary, as feminist and postcolonial science and technology theorists have pointed out. No matter how effective other cultures' (including "women's cultures'") knowledge traditions are, were, or might have been for enabling effective interaction with natural worlds, they are not counted as real sciences. No matter how much modern sciences might have incorporated elements of other cultures' concepts and theories about nature, their mathematical and empirical techniques, and even whole bodies of their accumulated navigational, medical, pharmacological, climatological, agricultural, manufacturing, or other effective knowledge enabling prediction, and control of nature, these other bodies of knowledge are not counted as "real science" until incorporated into European knowledge systems. And no matter how poor at explanation, prediction, and control European sciences are—for example, with respect to social causes of environmental destruction, or the causes of patterns of carcinogens or contagious diseases—these inadequacies do not count against European sciences' purportedly unique universal validity. Such issues were explored in earlier chapters.[3] How is this pattern of prevailing thought to be explained in light of the fact that success at prediction and control of nature are always claimed to be the most reliable hallmarks of a real science?

When assumptions are used to define the unique identity and value of a people, its civilization, and its history in the way the universal science ideal and its supporting internalist epistemology tradition apparently do, critical examination of them obviously is going to draw forth deep psychic anxieties and resistances. When assumptions carry this much moral and political weight, exposing them to contrary evidence is not going to be a comfortable process.

Four unity-of-science claims. In the early twentieth century, the unity of science thesis became an important form in which the universality claim was widely defended. This thesis overtly makes three claims: there exists just one world, one and only one possible true account of it ("one truth"), and one unique science that can piece together the one account that will accurately reflect the truth about that one world. What this thesis means—what methodological, metaphysical, or other features could constitute the unity of physics, chemistry, biology, psychology, economics, and so on—are issues about which philosophers and scientists have been concerned to make sense.[4] However, as political theorists are aware, a fourth claim is also assumed in these universality/unity arguments: that there is a distinctive universal human "class"—some distinctive group of humans—who should be taken as exemplars of the uniquely or admirably human to whom the truth about the world could become evident. For early modern scientists and philosophers, such a group was those members of the new educated classes whose minds were trained to reflect the order of nature that God's mind had created, as God's mind had also created human minds "in his own image." God's mind, human minds, and nature's order were assumed to be congruent or homologous. Scientists and the educated classes that could see the truth and importance of scientific accounts represented the universal class that could learn to detect the one possible true account of nature's order. For nineteenth- and twentieth-century Marxists, the proletariat represented this universal class. This class alone, since its labor transformed nature into provisions for everyday human life to exist, had the potential to become the unique representative of distinctively human knowers. This class alone had the potential to detect the real relations of nature and social life beneath the distorted appearances produced by class society. Some forms of feminism have flirted with a similar kind of transvaluation of gender that considers the possibility that women are the uniquely human gender: if it made sense in sexist society to imagine men as the model of the uniquely human, then perhaps it is reasonable to consider how in many respects women's characteristics—their claimed altruism, pacifism, sensitivity to others' needs, or some other putative virtue—are more reasonably regarded as uniquely valuable models of the human, capable of producing less distorted understandings of natural and social orders. And some African Americans have claimed that the suffering, compassion, or some other characteristic of African Americans under the horrible conditions of slavery uniquely equips them to understand natural and social orders in ways unavailable to those who have not had such experiences.

There are important insights behind such claims.[5] In the contemporary world of multicultural, postcolonial, and (more complex and diverse) feminist politics and social theories, however, faith has declined in the possibility and desirability of such a universal class—Enlightenment, proletarian, feminine, or culturally distinctive in any other way. In these worlds in which we all live (whether or not we acknowledge the effects post–World War II emancipatory social movements have had), who could such a distinctively human, universal class be? What group could democratically gain assent to their own abilities uniquely to represent accurately universal human interests and the one true natural and social order such interests supposedly could reveal in the face of other groups' different but also valuable cultural conditions for the production of knowledge for them and their survival? In contemporary life, many kinds of important differences between humans—biological and, more importantly, social, economic, political, psychic, and otherwise cultural—are recognized as resources for producing effective knowledge and advancing democratic social relations.[6]

The universality/unity ideal is no mere philosopher's notion; in one form or another it has been one of the most central and enduring values of otherwise conflicting conceptual and political tendencies in modernity's social theories. However, it is now attracting critical attention from many groups around the globe which claim that for them it has had primarily bad scientific, epistemological, and political effects.

2. Science and democracy: Allies or enemies? Does the ideal of a single, universally valid science decrease global democracy? Does the goal of global democracy advance or obstruct the plausibility of such an ideal? In the late nineteenth and early twentieth centuries, defenders of the universality ideal hoped that it could serve as a powerful antidote to the tides of racialist and nationalist partisan conflict that again and again had resulted in violence and even genocide. For them, appreciation of the universality of science and its standards of rationality and objectivity could only support and advance democratic social relations. Today the universality ideal's defenders see in the unique standards of scientific rationality and objectivity the main hope for restoring what they think of as the fair and orderly social relations now being disrupted by the claims and demands of multiculturalism, feminism and "relativism" in the post-Kuhnian social studies of science. In effect, these defenders of the universality ideal fear and do not find plausible the analyses produced by the three distinctive schools of science and technology studies that have emerged since World War II, and whose arguments have been examined in earlier chapters.

For many feminist, multiculturalist, and post-Kuhnian science studies theorists, however, the universality ideal increasingly appears as a force for maintaining inequality and obstructing democratic tendencies, and for obstructing the growth of knowledge. For these groups, claims to the transcultural truth of modern sciences' representations of nature, and of only those of modern science, function to mask the ways that modern sciences and their representations of nature's order tend to distribute the cognitive and social benefits of scientific and technological changes disproportionately to those already positioned to take advantage of them, and the costs primarily to those least able to resist them. Moreover, universality claims legitimate the devaluation and even destruction of knowledge traditions that have enabled women, the poor, and less powerful cultures to interact effectively with their environments. The unique universality claims also have bad epistemological and scientific effects, in addition to their political consequences. In a variety of ways, they function to increase the production of systematic ignorance. From this perspective the universality claims are epistemo-logically, scientifically, and politically dysfunctional.

Of course the familiar "universalist" response to such claims is to insist that the "anti-universalists" are confused. Lamentable as the worsening situation of women and racially and culturally disadvantaged groups may be ("if this *is* the situation," they demur), such arguments only address the applications and technologies of science, and nobody denies that these often are shaped by anti- as well as pro-democratic politics. Of course politics can misuse and abuse applications of sci-ences and technologies—"Think of Lysenkoism! Think of Nazi science!" they argue. Real sciences could not possibly have any political conse-quences. "Real sciences" simply provide pure information about nature's order, according to this older view. One can note that such a position also blocks the argument that science, its rationality and objectivity, support or advance democratic social relations, or have any other good social effects, however. If science were culturally neutral, then it could not have any social effects at all. Of course this recognition motivated the invention of "positivism" in Auguste Compte's proposal that the pure information that sciences produce is politically neutral, and the only *positive* social effects of science are to be found in its distinctive method.

However, this attempt to disassociate modern sciences from their effects has become increasingly difficult to defend. Earlier chapters reviewed the evidence against this "pure science" claim when they focussed on the explicit mission-directed character of so much valuable scientific research, from Galileo's work in the Venice armory and Pas-teur's concerns with public health, to contemporary medical, economic, and military-directed scientific and technological projects.[7] Such histo-ries clarify that mission-directed research should not always be concep-tualized as an obstacle to the growth of scientific knowledge; obviously, it can produce valuable information about and explanations of nature's regularities, whether or not one approves of the "missions." Sometimes the cultural interests and values that constitute a scientific project, its conceptual framework, methods, and purpose are politically relatively uncontroversial; sometimes not. After all, medical research to discover the causes of cancer or of AIDS, and research to establish space satellites for military surveillance have all produced reliable and valuable infor-mation about nature's regularities regardless of how one evaluates the social desirability of such projects. How are the technologies and ap-plications of a science to be regarded as completely separate from that science's information when the information is produced specifically *for* such technologies and applications? Obviously, to start with, the pat-terns of knowledge and ignorance that a science or collection of sciences generates bear a close relation to the culturally local overt purposes and unarticulated interests in such information. If such science is conceptualized at its cognitive core in ways suitable to culturally local medical or military purposes—whether or not individual scientists are aware of such a match—in what sense is it "pure"?

Another kind of evidence against the purity-of-science thesis, however, showed that technologies of scientific research themselves contribute to constituting what can count as legitimate scientific knowledge, and that those research technologies themselves have always been constituted through social, economic, and political processes that have social, economic, and political consequences. Such research technologies are part of scientific methods, yet modern science's methods were long cited as value neutral and thus responsible for science's unique universality. Consequently, philosophers of science have been forced to reexamine what it is about scientific methods that makes modern science so effective if it is not their social and political neutrality.[8]

Thus, the assumption that the universality ideal can only advance both the growth of knowledge and a democratic social order has come under increased suspicion from a variety of sources.

3. *The epistemological status of universality/unity claims.* The universality/unity claims express ideals, not the results of empirical observation. The question of this chapter is, therefore, whether they should retain their status as ideals. However, many of their defenders seem to think they are scientific claims; that the history and present achievements of science somehow prove that it is uniquely universally valid and unified.

One stream of common sense captured by the older histories and philosophies of science seems to conjoin everyday experience and historical evidence to make the universality thesis appear obviously supported by empirical evidence. In this way of thinking, our observing, perceiving bodies-with-rational-minds and their environments seem to fit together in ways that enable modern sciences to gain greater and greater accuracy in their predictions and control of nature. How else could we account for the many magnificent achievements of modern medicine, the ability of humans to walk on the moon, or the possibility of composing these sentences on this computer? Modern sciences must be providing more and more pieces of the one coherent account that reflects or uniquely corresponds to how the world is for these technological achievements to be possible. Doesn't the history of such successes constitute incontrovertible evidence that there is one nature, one truth about it, and one science that captures that truth? How could scientific predictions increasingly achieve such accuracy were this not the case?

On the other hand, before the pull of this way of thinking becomes too irresistible, we can recollect that common sense and the history of science also tell us that there is something wrong with it. We all know that scientific claims can never be regarded as once and for all proved (or disproved). They always must be left only tentatively confirmed by observation and reasoning since new evidence continually shows how familiar scientific ways of thinking have limitations that were not earlier visible. Every scientific project makes "background assumptions"about properties of the instruments, theories of vision, what the relevant variable local conditions are, the cultural neutrality of the relevant conceptual framework, and much more that can at another time be thrown into doubt. Perhaps even more importantly, any given scientific framework eventually outlives its usefulness in advancing the growth of knowledge. At moments of revolutionary scientific change, when a new framework promises to replace the older one, old data is repositioned within a different conceptual scheme. Most of the observations made in Ptolemaic astronomy were repositioned within Copernican astronomy. Do they provide "the same information" within these conflicting conceptual frameworks? Well, yes and no. Of course many observations of the moon, the sun, and the planets made prior to Copernicus and Galileo were retained in the heliocentric account. In one sense we see "the same heavens" that cultures did a millennium and more ago. On the other hand, what patterns such observations form, how such patterns are to be explained, and what such observations, patterns, and explanations mean to different groups of scientists and their diverse audiences (intended and unintended)—these differ immensely between the two theories. In important respects, the pre-Galilean observations do not provide "the same information" for heliocentric theories.[9]

In recognizing the importance of this kind of understanding of the history of science, one need not hold that new conceptual frameworks are fully incommensurable with the older ones, or that ones from different disciplines leave their practitioners unable to communicate with each other or to work together on scientific projects, as both "incommensurabilists" such as Thomas Kuhn and also his critics claimed must be the case when one gives up these kinds of universalist assumptions. Temporary, local *universalizing* strategies are devised in the face of the de facto unavailability of reliable universality claims. Historical and sociological evidence shows how scientists continually make effective, "good-enough" translations—pidgin languages—and technical equivalences to get from one conceptual terrain to another and to enable them to work together effectively. Thus, their joint scientific projects can draw on otherwise disunified, heterogeneous, local scientific practices and cultures. For example, Peter Galison examines the "trading zones" scientists create between their otherwise disunified work. When H-bomb designers, logicians, aerodynamical engineers, and statisticians sat down together in the 1940s and 1950s to construct computer-simulated realities, they brought to such interactions very different notions of "randomness," "experiment," and other terms.

> While the mathematician thinks about the best definition of "random" quite differently from the physicist, in the cauldron of those early days of computer simulations a notion of "random enough for present purposes" emerged, borrowing from several cultures, yet belonging exclusively to none.[10]

Others have pointed to the diversity of such translation devices developed not only in modern science traditions but also in the scientific and technological traditions of other cultures. Modern sciences are like other scientific and technological traditions in that they have developed just such means of establishing effective continuity in scientific and technological projects by peoples with disparate, heterogeneous kinds of knowledge. Historians, philosophers, and ethnographers of sciences and technologies are still struggling to come up with a way to represent adequately the immense innovativeness of scientific and technological workers in developing these complexes of conceptual, technical, institutional, rhetorical, practical, and other kinds of devices.[11]

Such analyses of the strategies scientists use to communicate and work together across their heterogeneous cultures have not yet come to inform common sense. Nevertheless, reflection enables anyone to realize that in order for the growth of scientific knowledge to have occurred, scientific change must be more open to the value of alternative models of nature's order than the "one world, one truth, one science" ideal (and one culturally distinctive kind of ideal knower) can suggest. The ideal could not accurately be reflecting the history of science as that is widely understood after more than three decades of the post-Kuhnian, feminist, and postcolonial science studies. These schools of science studies delineated the resources provided for the growth of knowledge by assuming many human worlds, many truths, many sciences, and many culturally diverse knowers.

Thus, the universal science ideal appears to be inconsistent with the best history of science and with the best contemporary scientific practices. If the results of scientific research cannot and should not be protected from historical change, then it is not clear what it could mean to claim them to be universally valid, let alone uniquely so. On the other hand, before leaping gleefully to the presumed final defeat of this particular conceptual framework and the social programs in which it was embedded, we must note that the defenses of the localness, heterogeneity, and disunity of science also should not be taken to be transcultural truths, fully supported by incontrovertible evidence from within the sciences and from the history and sociology of science. The argument here, instead, is only that the universality arguments seem to block our understanding of the history and necessary practices of the sciences. One could say that both the universality and anti-universality arguments are strategies for keeping understandings of human knowledge of nature in balance when confronted with tides that threaten to pull too strongly toward insistence either on unique universality or only on incommensurable localness in knowledge systems.

There is another kind of defense of the unique universality ideal that must be considered. Are there weaker forms of it that might be more plausibly defended?

Weak forms of the universality claims. i. Cultural plurality of scientists attests to universal validity of modern science. Let us set aside first a popular defense of the universality thesis that consists in pointing to how modern science's creators and users today come from many conflicting cultures, yet all agree to scientific claims. The individuals who have made modern sciences have come from Great Britain, Japan, Germany, India, Denmark, and many other nations; they have held diverse religious, political, and other kinds of cultural beliefs; yet they have been able to agree to scientific claims though there might well be little else to which they could agree. Something about modern science must be universally valid for it to emerge from such culturally diverse peoples in the first place, and then to continue to gain assent from more and more culturally diverse peoples around the globe.

This popular argument must be taken seriously since it is just the one that the Vienna Circle scientist/philosophers made in other forms: in a world of partisan conflict, Rudolf Carnap and others argued in Europe in the late 1930s, pursuit of the universally valid standards of scientific rationality provided the only imaginable hope for achieving peaceful resolutions to social problems. From this perspective, Nazism and the Holocaust were the consequence of too little scientific rationality. Such icons as this argument make it difficult to resist the appeal of "one nature, one truth, one science."

One can gain a useful perspective on such an issue by noting that European scientists have all agreed to claims that were, and for many people from other cultures still are, embedded in other, non-European worldviews. The number "0," to take one example, has distinctive cultural meanings for many peoples. It represents the beginning or origin of all that follows. From nothing emerges everything. Thus, it is often regarded as different in quality from all the numbers that follow. It is not in the same cultural category as 1, 2, 3 and the others. Indeed, the number one often has such distinctive properties in many cultures: from homogenous unity emerges the heterogeneous multiplicity of the world we know. Moreover, zero has been invented or discovered independently in several different cultures, in each of which it has occupied a distinctive place in the prevailing cultural discourses. However, it, like all the other numbers, is presumed in modern scientific thought to have no culturally distinctive meanings at all, though even in European history prior to the modern era zero also carried different meanings for different groups. Is the cultural neutrality presumed for numbers in the modern European worldview itself a culturally distinctive feature of that worldview? Chapter 4 examined why it is illuminating to understand how such assumptions and ideals of cultural neutrality are themselves culturally distinctive.[12]

To take another example, now acupuncture has been integrated into European medical practices for the control of chronic pain. Of course what has been extracted from the Asian systems of thought and practice from which it has been borrowed is just those elements that will fit into modern biomedicine. Discarded are the rest of the Asian beliefs about bodies that get in or out of "balance," and how to keep them in balance, that do not coherently fit into modern biomedicine's ontology and epistemology.[13] Yet we can ask: if these elements of other cultures' knowledge systems have become scientific once they have been adopted into the culturally distinctive ontology, epistemology, psychic structures, and political economy of European sciences, why weren't they just as scientific before, when they were embedded in different culturally distinctive interests, discourses, and ways of organizing the production of knowledge? After all, zero and acupuncture "worked" both theoretically and practically in those contexts no less than they do in this one. And why should we regard those larger bodies of culturally distinctive beliefs of other cultures as only containing obstacles to the growth of knowledge about the natural world when those cultural values and interests produced the thinking and practices of zero and acupuncture that have proved so valuable far beyond their cultures of origin?

Such reflections suggest that we need a different way of thinking about the fact that the claims of modern sciences, like those of other cultures, "hold true"—or, can be useful—far distant from their sites of original observation, and for phenomena that may be described and explained very differently in other cultures. They enable accurate prediction and control of nature. Evidently claims emerging from other knowledge traditions also can be regarded as universal or universally valid in this sense no less than modern scientific claims. And we can ask how such knowledge is different from the ways Ptolemaic astronomy and Aristotelian physics, in spite of their culturally local meanings and referents, predict accurately a great deal that Copernican astronomy and Newtonian physics respectively also predict. Don't modern sciences in such cases agree to the claims of knowledge traditions that are not modern European ones? It no longer appears reasonable to argue that while some cultures manage to stumble upon empirically adequate claims, only modern science's claims are produced by culturally neutral, and thus universally valid, methods.

ii. Metaphysical, methodological, and reasoning unity. Attempts to support the unity-of-science thesis and its universality ideal have been worked out in other forms by philosophers of science. John Dewey meant by the unity of science only a "scientific attitude"—a weak form of only methodological unity. For Rudolf Carnap, however, it was a linguistic unity that was important. The laws of biology and physics both "could be expressed in terms of everyday physical terms and procedures."[14] He did not mean that physical laws could replace biological ones, or that the entities of physics and biology would turn out to be one and the same. Otto Neurath's understanding of what the "unification" of science entailed was much like Carnap's.

On the other hand, Victor Lenzen thought that the laws of different fields of physics, and of physics and biology could be integrated. To Hilary Putnam and Paul Oppenheim, the domains of science could be reduced one to another, physics forming a base for the familiar hierarchically structured pyramid of unified sciences. And in the 1990s, physicist Steven Weinberg argued for a metaphysical unity of science that justifies U.S. government funding of the superconducting super collider (in his home state of Texas). Such a project would enable modern science finally to identify the fundamental constituents of the universe.[15] Such a metaphysical unity of the sciences produces "fundamental facts" that must be part of any subsequent science worthy of the name.[16]

Philosopher Ian Hacking points out that the idea of unity has always emphasized or blended two distinct ideas: singleness and integrated harmony. These have been weighted in different proportions in defense of diverse kinds of metaphysical unity, methodological unity, and logical or "styles of reasoning" unities. Here is where the unique universality ideal (singleness) can be seen to be embedded in the unity (integrated harmony) arguments. Surveying diverse arguments for just metaphysical and methodological unities, Hacking identifies "a metaphysical sentiment, three metaphysical theses, three practical precepts, and two logical maximums" each of which weights differently unity as singleness and unity as harmony.[17] Moreover, there evidently are as many styles of scientific reasoning as there are of effective human reasoning more generally. Crombie identifies six important ones, each of which brings its own standards of adequacy.[18]

Thus, these post-Kuhnian science and technology studies arrive at the same assessment that the postcolonial and feminist science theorists do: science cannot plausibly be understood as one single kind of thing at all. "There is no set of features peculiar to all the sciences, and possessed only by sciences. There is no necessary and sufficient condition for being a science."[19] In the face of such disunity, many different kinds of techniques serve as "unifiers," as Hacking refers to them. Mathematics is the earliest recognized to have such a function. However, it turns out that there is no "one thing" that is mathematics since it, like the sciences to which it gives an appearance of unity, is a motley collection of principles and practices, as he and other historians of mathematics point out.[20] And numerous other such unifiers are to be found among scientific instruments, techniques, attitudes . . . all those inventive strategies that occur in the "trading zones" within which scientists in the modern West, like other indigenous knowers, work to communicate across the diverse cultural and natural conditions that separate them.

Thus, common sense, the history of science, and observation of contemporary scientific practice do not support either the inherent singleness or "integrated harmony" of modern science. Moreover, earlier chapters explored the scientific, epistemological, and political advantages of understanding science as necessarily and desirably plural. Let us briefly review those findings and then turn to consider some techniques pointed out in the postcolonial and feminist accounts through which the illusion of the unity and universality of modern sciences has been historically established.

4. *Producing illusions of unity and of universality.* Some of the most important cognitive, technical resources of modern sciences are to be found in their distinctively local features, as post-Kuhnian, postcolonial and feminist science studies showed in earlier chapters. Of course nature's order—"reality"—has a great deal to do with what sciences— modern or other—come up with as the best results of research. "Nature" is a major player in producing sciences' most reliable and widely accepted claims, as is the case also with much of the belief of everyday life. People who do not pay attention to known regularities of nature, or who are disinterested in charting what turn out to be important regularities of nature, are among those who tend to live shorter lives as they are felled by avoidable "accidents" and other threats to life and limb. But nature does not have *everything* to do with even the best representations of such regularities. Scientific claims are not mere reflections of nature's order such that no traces of social values, interests, and inquiry processes are visible in such claims—let alone in the patterns of such claims that the sciences of different eras and different cultures tend to produce. As earlier chapters explored, it turns out that nothing in science can be protected from cultural influence—not its methods, its research technologies, its conceptions of nature's fundamental ordering principles, its other concepts, metaphors, models, narrative structures, or even formal languages that all play crucial roles in advancing the growth of knowledge—that is, its research questions and the consequent distinctive patterns of systematic knowledge and systematic ignorance that it generates. Many socially constituted theories about nature can be *consistent* with nature's order, but none are uniquely *congruent* with it; none uniquely correspond to it. Indeed, we should not want to protect sciences from all cultural influences, for many of these are precisely re-

sponsible for their great successes and are necessary for their continued successes in the future.[21] This is where the universality ideal turns out to be costly to the growth of scientific knowledge.

Indeed, it was precisely the use of local resources that enabled modern sciences to emerge in Europe according to postcolonial accounts. European expansion and the growth of modern sciences in Europe were causally linked in that each contributed important resources to the success of the other. Without the knowledge of those daunting aspects of nature's order that Europeans encountered in their "voyages of discovery," knowledge provided by the emerging modern sciences, Europe could not have successfully developed the imperial and colonial relations that permitted it to achieve global leadership. Modern sciences helped Europe shift from being just one of a number of cultures around the globe that were living through the beginning of the end of feudalism in the late Middle Ages to becoming the single most important center of the early or proto-capitalist global economic and political relations that it would spread around the globe. Moreover, without the diverse resources provided by European expansion, modern sciences would have had a far more difficult time emerging—perhaps they would not have emerged in Europe.[22]

Through this process, modern sciences developed distinctive patterns of systematic knowledge and systematic ignorance, traces of which remain visible in contemporary modern science. Early modern science produced information about those aspects of nature's order that European societies needed in order successfully to expand into the Americas, around the Cape of Good Hope, into Asia and eventually Australia, New Zealand, and Africa (and, now, out into space stations and onto other planets in our solar system). As some of the development theorists have argued, the "science and technology transfer" involved in post–World War II so-called development projects that overtly were intended to bring the "underdeveloped countries" up to the standard of living of the "developed countries" in fact have continued the colonial process begun five centuries earlier. These expansionist projects shaped what modern sciences would and would not know about nature's order.[23]

Chapter 4 identified four respects in which this kind of postcolonial account as well as the post-Kuhnian and feminist accounts argue for the advantages as well as the limitations that local resources provide for the growth of scientific and technological knowledge for every culture's knowledge projects. There it was pointed out that different cultures are located in different parts of heterogeneous nature's order; their environments are always local ones whether restricted to a Pacific isle or the trajectory between Spain and the Caribbean, or Cape Canaveral and the moon. Cultures are interested in whatever they count as their own environments, but even in "the same" environment, they will tend to have different interests generating different questions about the world around them. For example, on the shores of the Atlantic, one culture will be interested in fishing, another in coastal trading, a third in oil and minerals lying beneath the ocean's floor, a fourth in possibilities for transporting slaves, sugar, and rum back and forth across it, a fifth in using the ocean as a toxic dumping site, and so forth. Such culturally local interests lead to different patterns of knowledge and ignorance about local environments.

Moreover these patterns are organized and produced through culturally distinctive discursive resources. Both advancing and limiting a culture's patterns of knowledge are metaphors, models, and narratives about, for example, the Garden of Eden, peaceable kingdoms, wild and unruly nature, nature as a machine, or as a product of God's mind, as a computer, a spaceship, and a lifeboat; about noble, innocent, childlike, animal-like, primitive, or evil natives; about manly, heroic explorers, conquistadors, and natural philosophers, or ones that purportedly represented the admirable national temperaments of Spain, England, France, or other European nations. These kinds of discursive resources represent a distinctive European legacy rather than, for instance, one that could be found in an Islamic or Native American culture. Finally, the production of knowledge is organized in the distinctive ways that different cultures tend to organize social activities more generally. The characteristic ways that work, travel, conquest, and other social relations were organized in fifteenth- through twentieth-century societies of Europe and its diasporas shaped how the work of science and of European expansion were organized. The "voyages of discovery" were distinctively European ways of organizing the production of these parts of modern sciences' knowledge. Though these four kinds of local resources have been described here as if they were completely separate, in daily practice they are partly interlocked and shape each other.

Similarly, as earlier chapters explored, post-Kuhnian and feminist histories, sociologies, ethnographies, and cultural studies of modern sciences produced in the last thirty years have charted precisely the ways that scientists have used local resources to generate such new theories and interpretations. Thus, the postcolonial studies converge with northern science and technology studies, and with feminist components of both, in highlighting the strengths as well as the limitations of sciences' uses of local resources. Modern sciences are "local knowledge systems" no less than are the science and technology knowledge systems of other cultures. Of course modern sciences in many respects are much more powerful than other cultures' knowledge systems, though other cultures' knowledge systems also have their relative strengths over modern sciences, for their locations in nature, interests, discursive resources, and ways of organizing the production of knowledge enable them to learn patterns in nature's order that are not visible from modern science's perspective. But no matter how global the successes of modern science's predictions of nature's order, they can never achieve a unique universality in the sense of being culture free, or destined to persist through history with their meanings and conceptual contexts unchanged. Further, significant parts of the knowledge systems of other cultures also have been able to achieve effective prediction far from the original cultural location of their production.

If it is not useful to think of modern sciences as single or harmoniously integrated in the various senses reviewed above, how is it that we have all been fooled about this? Noted above were various strategies of translation, "trading zones," mathematical and other unifiers that have permitted modern scientists to think and communicate across their disparate projects. Postcolonial histories of science and technology enable us to identify several practices of European expansion that have also contributed to this effect. First, as such expansion turned the world into a laboratory for emerging European sciences, Europeans could test the hypotheses they developed over vastly larger and more diverse natural terrains than could other cultures. European expansion gained access for European sciences to a far greater diversity in nature's order than was available to cultures not so engaged in expansion and, in some cases, whose trade routes and other travels Europeans curtailed. Not all the sciences benefitted equally from this aspect of European expansion, of course. Astronomy, physics, and chemistry benefitted less than did cartography, geology, geography, climatology, and many kinds of biology, though they still did benefit as they addressed the challenges of expansion and its effects on European economies, and, more indirectly, as they became funded through riches gained from the Americas.[24] Agricultural, pharmacological, and medical sciences immensely benefitted from expansion, as did such social sciences as linguistics and anthropology. Of course any culture engaged in such expansionist projects could also have developed their systematic knowledge about natural and social worlds. Indeed, they would have to have done so in order to succeed at travelling through or settling in unfamiliar environments and climates, and cohabiting with the indigenes they encountered. The internalist epistemology of modern science has provided no resources for understanding the effects on European sciences of these expansionist projects.

Second, European sciences could forage in other cultures for elements of those cultures' "ethnosciences" to incorporate into European sciences. It was not just "hypotheses" about nature's order that Europeans came up with all by themselves that were tested in the course of expansionist projects. Native informants taught Europeans about the local flora and fauna and how to use them, minerals and ores and how to extract them, climates and how to survive them, diseases and other threats to health and how to avoid them, pharmacological remedies, agricultural, fishing and engineering practices, land and sea routes, and much of the rest of the knowledge traditions developed and stored in local cultures. Incorporated into the European sciences were those parts of this local knowledge that fit into the prevailing European conceptual frameworks. Those parts that did not fit were ignored or rejected. As the historians point out, not all the resources for modern sciences that the Europeans encountered in other cultures' knowledge traditions were initially perceived to be such, however. Europeans encountered mathematical notions they could not use till decades and even centuries later, and pharmacological knowledge that only now are the northern pharmaceutical companies systematically interested in gathering.[25] A history of "unborrowed knowledge" can provide an illuminating accompaniment to the histories of borrowed knowledge.

Moreover, the Europeans could combine knowledge gathered through observation or foraging from one part of the globe with knowledge so gained elsewhere to create kinds of knowledge that could not emerge from fewer sites in nature and in culture. This, too, is a unifying strategy. For example, Linnaeus's categories were designed to accommodate species from many different parts of the world; Darwin's hypotheses came to him as a result of thinking back and forth between what he had learned at different sites in his travels. Conceptual frameworks designed to explain the relation between observations made at different sites around the globe contributed to the idea that universal sciences were in the making.

Third, at the same time, European expansion suppressed or destroyed—both intentionally and unintentionally—competitive local knowledge systems. Some cultures were wiped out by diseases inadvertently carried by the Europeans; they were infected before there was a chance for them to be conquered, as one historian puts the point.[26] Others were destroyed by conquest. In both cases, the cultures took to the grave their repositories of knowledge about nature and social relations.

Even when the indigenes survived their first encounters with Europeans, their local knowledge traditions were often destroyed nevertheless, both intentionally and unintentionally. For example, the British set out to destroy the Indian textile industry, and succeeded in doing so, in order to sell their British-made textiles in the Indian market. In the United States, Native Americans were neither permitted to speak their native languages in the government schools nor to develop their traditional repositories of knowledge there. Again, the British did not permit Indians to learn the mathematics that had been created by Indian mathematicians.[27] Land upon which local knowledge traditions depended was appropriated by Europeans, turned to "scientific" agriculture, forestry, or other profit production for Europeans, and often environmentally impoverished. In such ways the basis for local knowledge traditions was removed from local access and often destroyed.[28] So the suppression of other cultures' knowledge traditions also contributed to producing the illusion that only European sciences were and could be universal ones.

A fourth way the illusion of unique success was created has been through the dissemination of a predatory conceptual framework for and by European sciences. This conceptual framework spread through the societies Europeans encountered as a central feature of the imposition or adoption of European culture. What is meant by a "predatory conceptual framework"? One way this occurred was through the persistent substitution of abstract, transcultural and ahistorical concepts of nature and processes of gaining knowledge for concrete, locally situated, and historical ones. The former were claimed unique to modern sciences and responsible for their successes, and the latter devalued as merely characteristic of "folk science." For example, features of local environments become aspects of omnipresent "nature" to be explained adequately only by universally valid laws of nature.[29]

There is nothing wrong with abstractions and generalizations in themselves. The point is rather that such abstract concepts always must in fact be accompanied by local knowledge about how to apply such concepts—when and where they are relevant, how to revise and extend them. Yet such an abstract, universalizing conceptual framework devalues this very local knowledge that it needs in order to complete our understanding of it as empirical knowledge—how it relates to the world around us. One could say that the abstract and universal perpetually depend upon and reproduce the "premodern" forms of local knowledge required for the "universal" to be regarded as empirically relevant. It is not that modern science actually replaces its premodern predecessor; rather, it insists on its continual production as a devalued form of knowledge.[30] Moreover, such "foreign" concepts consistently have been used to legitimate the authority of powerful groups over economically and politically vulnerable ones.

Most effective in establishing the impression of universality is simply insisting upon it as an empirical fact: there is one and only one kind of "right" or "real" science, and that is the kind practiced in modern Europe. This replication in modern science of the monologic voice of the Judeo-Christian God is buttressed with various supporting theses—scientific accounts are value free, nature is value free, no kinds of interventions in or uses of nature are forbidden either by nature or by science, and so on. However, as reviewed above and in earlier chapters, such claims are not themselves the results of scientific or historical investigations, nor is there anything logically necessary about them. They are, instead, articles of faith, as is the insistence in modern science on its own "solo performances," that are so well suited to the belief in nature's "monovocality."

Thus, the appearance of universality is created not by any internal epistemic features of modern sciences, as the universality ideal assumes. Instead, it is produced by the kind of hard scientific work reported in the post-Kuhnian accounts and by contingencies of history and political strategies, obscured by, or fit into, a dogmatic conceptual framework that is persistently rhetorically elaborated. This will sound like a harsh judgment. Yet it is important to state as clearly as possible these findings of postcolonial science studies that are strongly supported also by the other two schools of science studies examined here. We need better ways to conceptualize the successes and limitations of modern sciences than are provided by the universality ideal of the internalist scientific epistemology. The ideal neither fits the facts of state-of-the-art history, sociology, and philosophy of science, nor does it make sense in light of what is now understood about how human knowledge of the nature's order must be gathered, preserved, and expanded.

5. The local-global continuum: An alternative conceptual frame. If the successes of sciences—modern or other—cannot be attributed to their internal epistemic features—such as a uniquely universally valid metaphysics, methodology, language, or standards of objectivity and rationality—what does account for them? Apparently there is no distinguishing feature of a science, as we saw earlier, but we still might usefully ask about causes of, or influences on, variations in the powers of different science traditions.

One strategy of many science and technology observers has been to try to bring into visibility the set of practices through which different modern scientific projects have maintained valuable tensions between the local and the global. Knowledge systems, any knowledge systems, always are constituted initially through a set of local conditions. However, the most widely successful ones, such as many parts of modern sciences, manage to travel effectively to become useful in other sets of local conditions—parts of nature, interests, discursive resources, ways of organizing the production of knowledge—that are different in significant respects from those that initially produced them. Without claiming a universality for them that we now can see is historically and conceptually misleading, how could we usefully think about valuable tensions between the local and this movability, or ability to travel, that has characterized parts of modern sciences in particular, but also parts of other knowledge systems (e.g., the concept zero and acupuncture)?

"Technoscience," proposed by Bruno Latour, is one term that has proved useful for drawing attention to the value of maintaining certain kinds of tensions between the local and the global in modern scientific practices.[31] Other science observers have focussed on different sets of components of such complexes that enable them to maintain the local/global tensions. As Helen Watson-Verran and David Turnbull point out, this is an area of study still emerging since no single analysis so far proposed quite captures all of the heterogeneous practices modern technosciences have developed.

> Though scientific culture is now being more frequently recognized as deeply heterogeneous (see, e.g., Law, 1991c; Pickering, 1992b), there is, at present, no term in general usage that adequately captures the amalgam of places, bodies, voices, skills, practices, technical devices, theories, social strategies, and collective work that together constitute techno-scientific knowledge/practices. Foucault's epistemes; Kuhn's paradigms; Callon, Law, and Latour's actor networks; Hacking's self-vindicating constellations; Fujimura and Star's standardized packages and boundary objects, and Knorr Cetina's reconfigurations—each embraces some of the range of possible components but none seems sufficiently all-encompassing.[32]

Watson-Verran and Turnbull's work is especially interesting because of the way it links European and other cultures' scientific practices in these respects. They are concerned to show how other scientific and technology traditions besides those of modern science have achieved similar kinds of balances between the local and the global in their most successful projects. They develop Deleuze and Guattari's term "assemblage" as a more satisfactory way to capture the set of techno-scientific "power practices" that enable cultures in different ways to maintain that crucial tension between the local and the global.[33] Watson-Verran and Turnbull show how medieval European cathedral builders, the Anasazi, the Inca, Australian aborigines, and Pacific navigators, like modern scientists, develop "social strategies and technical devices" that enable them to create "equivalences and connections whereby otherwise heterogeneous and isolated knowledges are enabled to move in space and time from the local site and moment of their production and application to other places and times."[34] This is not the place to explore their interesting account in greater detail. Rather, they provide a good example of the postcolonial arguments showing that it is not internal epistemological features but diverse combinations of technical and social strategies that enable both some modern and "indigenous" technoscientific traditions to become more successful than others. We do not need the notion of universal validity under consideration in this chapter, for other frameworks are becoming available that can preserve what was valuable in the universal science framework without the severe costs of the latter's historical and conceptual inadequacies.

It is time to summarize those costs for the sciences, for epistemologies, and for democratic social relations.

6. Dysfunctional universality claims: Scientific, epistemological, and political costs. To conclude, whatever the remaining benefits of supporting the universality ideal, the sources and arguments reviewed here show that its costs are significant. Let us look first at scientific and epistemological and then political costs of maintaining this ideal.

For one, the unique universality thesis supports the legitimacy of appeals to the authority of a single, monolithic "science" to support individual scientific claims, rather than each having to stand on its own—to "face the tribunal of observation" without the crutch of the general authority of modern science. Feminist and postcolonial critics have pointed this out again and again. Hypotheses about women's psychologies, reproductive systems, or physical abilities have achieved legitimacy when they can claim to be scientific ones, whether or not rigorous testing of such hypotheses has occurred.[35] The empirical reliability of agriculture and forestry principles developed for European environments has been presumed superior to local practices in African, Indian, and other environments when the former can claim the status of modern scientific principles rather than only the ones local farmers and peasants have developed and improved for generations.[36] In such cases individual scientific claims have not had to face the empirical tests that are demanded of claims that cannot appeal to modern science for their legitimacy.

This argument has been voiced far more widely. As philosopher John Dupre puts the point, "The political power of science rests in considerable part on the assumption that it is a unified whole." If science is disunified, then "particular appeals to the authority of science must stand on their own merits."[37] The unity-of-science thesis and its unique universality claim encourage what Dupre refers to as the "unity of scientism."[38] Without the crutch of such dogmas as the universality claim, many purportedly viable scientific claims would have to face much more rigorous tests of empirical and theoretical adequacy.

Secondly, the universality claim legitimates resistance to the most valuable criticisms of contemporary science. Feminist theorists frequently are challenged either to show the ideological bias in physics, or admit the irrelevance of any other kind of evidence to support their claims of androcentric assumptions in the constitution of scientific claims.[39] Criticisms of modern sciences that cannot be recognized as coming from within the sciences can be devalued or ignored without the kind of consideration that criticisms from within the sciences would receive by those who hold the unique universality ideal. Yet it is precisely the fact that they come from what is perceived to be outside the sciences that makes such critiques especially valuable. It is only by starting from outside the dominant conceptual frameworks that such frameworks can themselves come into sharp focus, as the arguments in earlier chapters for standpoint epistemologies and their standards for "strong objectivity" pointed out.

This issue has been central in the postcolonial accounts also. The global authority of a claimed uniquely unitary science, especially one associated with increasingly widespread eurocentric ideals of modernity, progress, unique human potential, and manliness defined in eurocentric terms, conspires to silence what are potentially the most viable alternatives to modern sciences' claims and concerns. The universality claim makes it difficult to see the limitations to modern sciences' institutions, cultures, and practices that accompany their strengths. Moreover, the universality claim works against the overt, valuable claim of modern sciences and their philosophies that it is vigorous criticism that most advances the growth of knowledge. Instead of encouraging such criticism, the universality thesis suppresses it. The way the universality ideal tends to immunize sciences against their most telling criticisms points to the problematic assumption of a single, unified "class" of knowers to whom responsibility for discovering nature's order should be assigned. Women, non-Europeans, and activists/scientists among such groups, who "bring their special-interest politics into science," have never been considered appropriate members of such a class of knowers by those who have benefitted most from scientific and technological change.

Third, the universality thesis is dysfunctional for the growth of scientific knowledge in another related respect. It has the effect of decreasing valuable forms of cognitive diversity, as the postcolonial critics in particular have argued. There is no evidence that the kinds of sciences favored in the modern North today will remain the most useful ones in the future either for other cultures or for the heirs of modern European cultures. Indeed, the arguments here, and many others explored earlier in this book, point to ways in which the ontologies and methodologies of modern sciences, and the interests and discursive resources that shape them, are not the most useful ones for many scientific research projects today. The universality ideal functions to delegitimate any but the scientific problems found interesting in the modern West.

Fourth, the strongest form of the universality ideal has raised distinctive obstacles to our understanding of certain kinds of ways the world is arranged and changes over time. It does this by promoting only narrow conceptions of both nature and science. As long as physics is assigned the status of the model for all sciences, whether on historical, ontological, methodological, logical, or other grounds, modern sciences will unnecessarily generate a certain kind of distinctive pattern of ignorance about the world around and in us. When physics, especially the narrowest conceptions of it, is permitted to set the standards for what counts as nature and what counts as scientific accounts, our knowledge will tend to focus disproportionately on discrete, isolated, short-term, and "purely physical" aspects of the world around us. (Here, the phrase "nature's order" starts to look suspiciously narrow.) It blocks our ability to get into focus the social elements—institutions, practices, languages, meanings—in what are often presented as purportedly merely natural, scientific, and technological changes. It makes it especially hard to see those that are distant, broadscale, and long-term.

Moreover, it blocks our ability to grasp systematic patterns of ignorance that any preferred pattern of knowledge will also generate. The universality ideal encourages the unfortunate tendency to internalize the benefits of scientific and technological change and externalize their costs. The benefits tend to be seen as the consequence of internal features of the epistemology of modern sciences, and the costs as the consequence only of misapplications of scientific knowledge or of their technologies, but not of scientific processes themselves.

Finally, such a model in the natural sciences also promotes the production of systematic ignorance in the social sciences. There are the social sciences that overtly model themselves on the natural sciences: physicalist psychologies, rational choice theorists in economics and international relations, and positivistic sociologies, for example. But there are also the social sciences that conceptualize their projects in such single-minded opposition to the naturalistic models prevailing in their research areas that they cannot get at the ways that more global forces shape or are the consequences of the social phenomena that they study. They get stuck in the local as a reaction to naturalistic social sciences' devaluations of the local. In such cases universalism's conceptual world is advanced in unarticulated forms.

Last but not least, there are the political costs. Feminist and postcolonial theorists especially have argued that the bad political effects of modern sciences' universality ideal are in part a consequence of the scientific and epistemological costs of this ideal. We have seen in this chapter and earlier ones how the universality thesis supports the devaluation of forms of knowledge-seeking that have proved valuable in non-western and premodern cultures, and in devalued subcultures in the West (and elsewhere), such as women's cultures. Indeed, modern sciences have ended up, unintentionally usually, partners with the worst genocidal social projects when they lend legitimation to the destruction of other peoples and the cultures that sustain them, not just their knowledge systems, as the inevitable costs of "human" progress. The centrality of European sciences and technologies to the further de-development of the world's least advantaged peoples in the name of human "development" is one place where such bad effects of the universality ideal can be seen. Here the universality thesis legitimates continuing to move access to nature's resources from those who are already the most economically and politically vulnerable to those who are already the best positioned to take advantage of such access. The universality thesis elevates to a desirable ideal models of the distinctively rational, progressive, civilized, and human that are constructed in opposition to, in terms of their distance from, the non-European, the economically frugal, and the feminine. Indeed, the universality thesis elevates authoritarianism—the necessity and desirability of acknowledging the legitimacy of just one true account of the world—to a necessity for the distinctively rational, progressive, civilized, and human.

The philosophies of modern sciences have always claimed that such modern knowledge-seeking contributes to democratic social relations. One can find such assessments throughout the evaluations of modern sciences from the Baconian New Science Movement in the seventeenth century through the advent of Comte's positivism in the nineteenth century, the logical positivist philosophies of the 1930s and 1940s, to today's debates about the appropriate projects for sciences and technologies after the Cold War. However, the explorations of the scientific and epistemological dysfunctionality of the universality thesis support long-voiced arguments that there is another, conflicting story to be told about the relationship between modern sciences and democratic social relations. Other conceptual frameworks can do the historical, empirical, and theoretical work that was provided by the universality ideal without invoking the latter's scientific and epistemological dysfunctionality or its ethnocentric, antidemocratic politics.[40]

11

Robust Reflexivity

Locating this study. What is the social location of this study—the one here drawing to a close? And what is its epistemological stance? Should this study pretend legitimately to escape the historical and epistemological questions it has asked of others?

One kind of answer to such questions was provided in the preface, where I traced my particular route to the issues of this project. My own history of teaching interests and scholarly projects, occurring in the contexts of local and global political shifts and of the social movements that have challenged and nourished me during the last quarter century, has shaped the analyses and discourses of this book in distinctive ways. Clearly, other contemporary science and technology researchers would find other issues more important and take different approaches to the issues of interest to me. Yet my own "personal history"—my intellectual/political biography—is not all that idiosyncratic, for the postcolonialism, feminism, and post-Kuhnian science and technology projects that have intrigued me in distinctive ways obviously have also interested others. And the kinds of approaches I have taken to them certainly are not unique to me. A "personal" response only begins to answer such questions. Alas—it may be that I can only partly be held responsible for the historical and epistemological positions this study occupies, whether readers want to praise or blame me for them! Alas, those obligatory disclaimers we characteristically produce—"I thank x, y and z for their helpful comments on this writing, but I alone retain responsibility for the claims made here"—such disclaimers may claim too much individual responsibility for the contents of our texts.

Thus, chapter 1 provided another kind of answer to the "location question." There, the advantages of creating spaces for more and richer encounters and exchanges between these three leading schools of post–World War II science and technology studies were proposed. In this sense, this study lies at the proposed convergence of northern post-Kuhnian science and technology studies, postcolonial science and technology studies, and feminist science and technology studies—this last appearing both in the North and in the southern "Women, Environment and Alternatives to Development" studies. (An equally reasonable way to mark just what is converging here would be to speak of the post-Kuhnian and postcolonial schools, each with a powerful—if not always fully appreciated or integrated—feminist component.) So the conjuncture of these leading schools of science and technology studies are significantly the site of this study, though other thinkers located there with me would think differently about the issues they situate there.

A related third kind of answer to the question of the location of this study was also provided in chapter 1, where central conceptual shifts that shaped this study's project were mapped in an attempt to make the narrative of this book less strange to readers unfamiliar with one or another of the literatures on which this study draws. These conceptual shifts seemed to me to be called for by findings of the three post–World War II science and technology studies. The conceptual field of the book used an expanded notion of science that set aside for the time its familiar contrast with ethnosciences or indigenous or local knowledge systems. And it found analyses of sciences' constitution by technologies of research and of application often crucial for understanding the cognitive core of a science. Moreover, the conceptual field of this study takes examinations of philosophic eurocentrism to be necessary to understand the history and philosophy of modern sciences and technologies, and of the scientific and technological traditions of other cultures. And it set out to try to imagine what kinds of epistemologies could better serve us in the kinds of intellectual and political worlds that post-Kuhnian, postcolonial, and feminist science and technology studies have revealed. Thus, it is located in an epistemological terrain that claims stronger standards for maximizing objectivity than the conventional philosophies of science have adopted, that uses existing power/knowledge relations to detect how such relations have shaped the conceptual frameworks and ensuing patterns of knowledge and ignorance of the science and technology traditions of the modern West and of other cultures, and that distances its project from the typical universality assumptions invoked to claim authority for our texts. So these are some additional features of the epistemological map on which this study is located.

Yet one more aspect of such a map was discussed in chapter 1. The categories and concepts invoked here have been proposed not to "name reality" in some authoritative manner but, rather, as strategic ones to be valued for their usefulness in enabling us to think about aspects of scientific and technological cultures—our own and others'—that would otherwise be hard to detect or articulate. The project here is to create spaces and resources for kinds of encounters and exchanges that are otherwise hard to come by. Their adequacy is, thus, a matter of their usefulness for such encounters and exchanges. We are tempted to think of maps as "reflecting reality" in some way or another. And they certainly bear determinate relations to the world around us, we hope. Yet we keep only the maps useful for the kinds of projects of interest to us. Of course they bear a "relation to reality," for if they mis-map where we want to go, we get lost and, if we survive our bad maps, discard them for more useful ones. But no single map or collection of them can perfectly reflect "reality." There is no one, perfect map that can chart everything I will ever want to know even about the short and relatively simple terrain that lies between my chair here in Santa Monica and the one at my desk in Westwood, for that terrain—the land, roads, climate, buildings, peoples, and social arrangements—itself changes continually due to earthquakes, floods, mudslides, droughts, immigration, emigration, Pacific Rim political economy, real estate speculation, outcomes of Los Angeles, California, national and international politics, and innumerable other aspects of the usual Los Angeles "changing of the seasons." And it alters as my interests, discourses, and ways of producing knowledge about it change. (See earlier chapters for discussions of these local constituents of patterns of knowledge and ignorance.) So the categories and concepts of the maps on which this study "travels" are strategic ones, not metaphysical ones in any conventional sense.

One might think no more could or need be said about the location of this study. Yet we can begin to explore its contours in other ways. One theme in this study has been to pursue the epistemological implications of understanding—that others can often identify better than can we—the conceptual frameworks that shape our beliefs and actions. "Would the gift the genie give us to see ourselves as others see us," as the old saying goes. We are the "natives" to our own institutional, social, and civilizational belief systems, and when these are systems shared with virtually everyone whom we encounter, we are even less likely to be able to detect the assumptions that shape the thought of all of us. Institutional, social, and civilizational or philosophic ethnocentrism is our inevitable lot. However, there are ways to reduce such forms of ethnocentrism, as earlier chapters have argued.

Another theme in this study has been that local aspects of modern sciences have not always been only the obstacles to the growth of knowledge that they have been imagined to be by conventional epistemologies. Local features of knowledge systems are not just "prison houses of belief" but also "toolboxes" that enable scientific and technological traditions to understand more about their natural and social environments and interact more effectively with them. What are some additional distinctive local resources and their limitations that have constituted this study?

Another local knowledge system. That these three schools of science studies, at the convergence of which this study lies, have emerged in such a way as to be visible now and here, when and where I am fortunate to be listening, watching, reading, writing, and conversing with individuals and groups in all three areas, is the result of social processes that historians and political economists can and will chart far more effectively than can I. From the perspective of most places in the world today, the trajectories of these three kinds of science projects would not converge and contrast in the ways I have described. Other themes would be more important at such locations. Using the book's conceptual map, one can already identify distinctive local characteristics of this study that will limit its usefulness in some quarters.

The standpoint epistemology recommended in earlier chapters, and its strong objectivity program, is indeed a "science project," first of all. This study joins others that intend to improve the performance of the sciences, natural and social. It argues that more illuminating and useful understandings of nature's order for improving the lot of the majority of the world's peoples can emerge only when a more adequate epistemology directs understandings of how sciences do and could produce knowledge. Thus, this study relegitimates modern sciences and attempts to rehabilitate some of their traditional central philosophic underpinnings such as conceptions of scientific rationality, objectivity, and method. However, it does so in a world where many think the power of modern scientific forms of rationality should be far more limited, not further expanded. Now the "context of discovery"—sciences' locations in the natural world, and their interests, discursive resources, and ways of producing knowledge—are to be added to the phenomena to be analyzed with this rehabilitated kind of scientific rationality, objectivity, and method. This problem can more fully be articulated as part of the next one.

This study's approach originates in the North, and at a particular moment and site of northern discourses. It draws upon the historical and cultural legacies of those cultures—for example, European and European feminist social theories, postcolonialism as articulated in a certain range of English-language writings, histories, sociologies, ethnographies, and philosophies of sciences as these are produced and debated in Europe and the United States, and so forth. Thinkers in other societies may well prefer to draw on the riches of their own cultural, intellectual legacies in order to develop resources for blocking "might makes right" in the realm of knowledge production. Why should they prefer yet one more theory that in its cultural specificity has the effect of devaluing their own cultural inheritances (regardless of how "postcolonial" I intend this study to be)? Surely northerners are not the only peoples to worry about local class relations, women's situations, or the global political economy and sciences' and technologies' roles in these. Standpoint theory's discursive tradition is undoubtedly a northern, European one; part of its strength in the North is created by the way it takes its places in older European class and feminist struggles, which have themselves included attempts to strengthen modern sciences, their ways of thinking and of organizing their production of knowledge. Its central assumptions, concepts, and narrative structures take their shapes in part through their fit both within and against other currents of northern intellectual and political history. Why should these great re-sources for standpoint approaches in the North be just as appealing to theorists from other cultures? There may, or may not, be valuable ele-ments in standpoint approaches for science and science studies projects originating in so-called developing societies. It would be astonishing and suspect, not to mention impossible, however, if this particular "pro-duction of knowledge" were equally useful to everyone, everywhere.

Moreover, it could appear to be socially located in yet another way. At a moment when conventional legitimations for northern scientific and epistemic authority are weakening, it can appear as yet another attempt to continue "imperialism by other means." Especially when part of its "object of study" is the knowledge tradition of southern cultures, this project can appear to try to appropriate and integrate into northern projects the important discourses of southern science critics. Our consciousnesses are not determined by our social locations, it argues; look at the ways that Europeans have helped to construct anti-imperialist and postcolonialist thought and institutions. Look at the men who have struggled on behalf of women's standpoints, who have actively interrogated masculine practices and cultures.

Standpoint approaches have what will appear to some as this lim-itation; they show how to think "from many social locations" that, in fact, northerners, or men, or literate elites in any part of the world, do not experience and live through. If one can think from these other social locations—men from the perspective of feminist discourses, women of European descent from the perspective of the lives of the women and men Europe has colonized—one might ask what is the difference be-tween standpoint approaches and "the view from nowhere"? Moreover, people from other cultures may not be all that interested in learning to think from the position of anyone and everyone in the world. Their struggle may be to figure out how to think most effectively from the standpoint of accounts of "their own" culture's history and present needs and desires that have been silenced and made invisible or close to invisible. After all, these have been the initial projects of northern feminist, multicultural, working-class, and lesbian and gay knowledge projects. Standpoint theory can appear to them to have only two trajec-tories open to it—identity politics and epistemology ("Only I and my kind can really know my/our worlds") or "imperialism by other means." Before finding oneself excessively tempted by the choice between identity and imperialism, we can note that this criticism does not make the other approaches standpoint theories were intended to outperform—ethnocentric ones, whether empiricist, interpretationist, or some other version of a "critical theory"—more attractive; they, too, are susceptible to just these sorts of criticisms in other forms.

Strategic robust reflexivity.[1] Can the standpoint approaches of this study be defended against such critiques? I think so; but only in the spirit of a tentative contribution to an ongoing discussion. Standpoint approaches still remain the best epistemology visible for northern sciences and science studies, one can argue. The global power of conventional science and technology projects and discourses, or of conventional histories and philosophies of science and technology, does not "wither away" by ignoring it. Standpoint approaches are, after all, critiques of the dominant power-knowledge relations. "Strong objectivity," with its enlarged understanding of research methods, the rejection of universality and unity-of-science assumptions, the kind of strategic "robust reflexivity" being explored here, and other epistemological features of this study may not be the crying need of every culture's theories of knowledge production. Yet, in affirming the intellectually and politically desirable potential of central European and North American epistemological concepts and hypotheses, they offer one set of possibilities for more realistic and politically progressive thinking about sciences and technologies in/and/of societies for observers and practitioners in the modern North. But to put the issue this way is to articulate an important limit of standpoint theories: they are useful as one counter to the dominant cultures and practices of science and science studies that emanate from the North. They may or may not provide resources for democracy-advancing projects in different cultural locations.

A robust reflexivity project conducted from a standpoint of a marginalized group will seek to detect the limitations, borders, edges of its cultural resources—where they thin out and run against other options. It will seek to avoid that only weakly reflexive position one can find in northern science studies that can only meet itself, mirrored back, as suffusing all of society and history. Some critics have argued that this weak reflexivity (as I am calling it) is characteristic of the problematic androcentric construction of self.[2] Postcolonial critics have identified it as distinctively eurocentric. Such forms of reflexivity are maintained by residual commitments to producing an internalist science of science, instead of a transformed local epistemological system for this particular time and place. Such weak reflexivity still is committed to the separation of "real science" from the large-scale political forces that shape sciences and science projects outside the control of individual researchers or small groups of them. And thus it is residually committed to the separation of the observer from the world he or she observes, for only science-studies observers, of all the objects in the universe, turn out to be able to provide "the view from nowhere" with respect to their locations.

This study has taken what we could call a "strategically reflexive" position. The point of trying to appreciate the historical location of its conceptual framework and political commitments is to enable more modest and thus more effective deployment of its claims. I hope that it can contribute to a dialogue occurring around us about how we might best think about the multiplicity of local resources, and their potential for generating knowledge, that become available in more democratic social relations. Of course, what we will mean by this latter term will itself take shape through local discussion.

NOTES

Preface

1. *Discovering Reality: Feminist Perspectives on Epistemology, Metaphysics, Methodology and Philosophy of Science* (Dordrecht: Reidel/Kluwer), which I coedited with the late Merrill Hintikka, appeared in 1983. In 1982 and again in 1983 I taught a course on African and African American Philosophy, and in 1982 I had organized a lecture series on black philosophy. Since the early 1970s I had been working through in my philosophy of social science courses issues about the rationality of other cultures and objectivity of this culture's observations of them as these appeared in the anthropology literature of the period.

2. Just a few of these scholars and some of their work to which my work is deeply indebted (whether or not they agree with the directions the latter has taken) are: Anne Fausto-Sterling, *Myths of Gender: Biological Theories about Women and Men* (New York: Basic, 1985); Donna Haraway, *Primate Visions: Gender, Race, and Nature in the World of Modern Science* (New York: Routledge, 1989), and *Simians, Cyborgs, and Women: The Reinvention of Nature* (New York: Routledge, 1991); Nancy Hartsock, "The Feminist Standpoint: Developing the Ground for a Specifically Feminist Historical Materialism," in *Discovering Reality*, ed. Sandra Harding and Merrill Hintikka; Ruth Hubbard, *The Politics of Women's Biology* (New Brunswick: Rutgers University Press, 1990); Evelyn Fox Keller, *Reflections on Gender and Science* (New Haven: Yale University Press, 1984), and *Secrets of Life, Secrets of Death: Essays on Language, Gender, and Science* (New York: Routledge, 1992); Alison Jaggar, *Feminist Politics and Human Nature* (Totowa, N.J.: Rowman and Allenheld, 1983); Genevieve Lloyd, *The Man of Reason: "Male" and "Female" in Western Philosophy* (Minneapolis: University of Minnesota Press, 1984); Carolyn Merchant, *The Death of Nature: Women, Ecology, and the Scientific Revolution* (New York: Harper and Row, 1980); and Dorothy Smith, *The Everyday World as Problematic: A Sociology for Women* (Boston: Northeastern University Press, 1987), and *The Conceptual Practices of Power: A Feminist Sociology of Knowledge* (Boston: Northeastern University Press, 1990). Of course there have been many other feminist scholars in many disciplines whose insights have been significant to my thinking then and now. Many of them will be cited in the essays that follow.

3. Thomas S. Kuhn, *The Structure of Scientific Revolutions*, 2d ed. (Chicago: University of Chicago Press, 1970). My otherwise wise philosophy of science teacher in graduate school in the late 1960s instructed us not to waste our precious reading time on this then fashionable book; the attention it was receiving was only that of a fad, and it would pass.

4. Especially powerful to me were Barry Barnes, *Interests and the Growth of Knowledge* (Boston: Routledge and Kegan Paul, 1977); David Bloor, *Knowledge and Social Imagery* (London: Routledge and Kegan Paul, 1977); and Bruno Latour and Steve Woolgar, *Laboratory Life: The Social Construction of Scientific Facts* (Beverly Hills, Calif., 1979).

5. My dissertation was on W. V. O. Quine's epistemology. Among the other philosophers whose writings I found especially powerful at the time were Paul Feyerabend, for example his *Against Method* (London: New Left, 1975), which I had seen in manuscript, and Richard Rorty, especially his *Philosophy and the Mirror of Nature* (Princeton: Princeton University Press, 1979).

6. A striking exception here was Donna Haraway's work on primatology, which took her to the African Hall of the Museum of Natural History and thence in her analyses to Africa and eventually Japan and India. See her *Primate Visions* and *Simians*, both of which include essays written earlier. See also Anne Fausto-Sterling's and Ruth Hubbard's work cited in note 2 for other northern feminist authors who took up race and postcolonial issues about science and technology early on.

7. Sandra Harding, ed., *The "Racial" Economy of Science: Toward a Democratic Future* (Bloomington: Indiana University Press, 1993).

8. See, e.g., Sandra Harding and Elizabeth McGregor, "The Gender Dimension of Science and Technology," in *UNESCO World Science Report 1996*, ed. Howard J. Moore (Paris: UNESCO, 1996).

9. See my "Are Truth Claims Dysfunctional?" in *Philosophy of Language: The Big Questions*, ed. Andrea Nye (New York: Blackwell, 1998).

1. A Role for Postcolonial Histories of Science in Theories of Knowledge?

1. Thomas S. Kuhn, *The Structure of Scientific Revolutions*, 2d ed. (Chicago: University of Chicago Press, 1970), 9.

2. Ibid., 1.

3. Ian Hacking identifies these two components of the unity of science thesis—the singularity and harmonious integration of scientific projects—in his "The Disunities of the Sciences," in *The Disunity of Science*, ed. Peter Galison and David J. Stump (Stanford: Stanford University Press, 1996).

4. Richard Rorty, *Philosophy and the Mirror of Nature* (Princeton: Princeton University Press, 1979).

5. See Peter Galison, "Introduction: The Context of Disunity," in *Disunity*, ed. Galison and Stump.

6. See John A. Schuster and Richard R. Yeo, eds., *The Politics and Rhetoric of Scientific Method: Historical Studies* (Dordrecht: Reidel, 1986), and, on the origins of experimental method, Steven Shapin and Simon Schaffer, *Leviathan and the Air Pump* (Princeton: Princeton University Press, 1985). This internalist epistemology is sometimes referred to as "positivist" precisely because of Comte's claim that the only positive contributions the sciences could make to society would arrive through the power of scientific method to produce accurate accounts of nature's order; otherwise, the sciences are culturally neutral.

7. See J. M. Blaut, *The Colonizer's Model of the World: Geographical Diffusionism and Eurocentric History* (New York: Guilford Press, 1993); J. M. Blaut, *1492: The Debate on Colonialism, Eurocentrism, and History* (Trenton, N.J.: Africa World Press, 1992); C. L. R. James, *The Black Jacobins: Toussaint L'Ouverture and the San Domingo Revolution* (London: Secker and Warburg, 1938); Eric Williams, *Capitalism and Slavery* (Chapel Hill: University of North Carolina Press, 1944); Ramkrishna Mukerjee, *The Rise and Fall of the British East India Company* (New York: Monthly Review Press, 1974).

8. See Andre Gunder Frank, *Capitalism and Underdevelopment in Latin America* (New York: Monthly Review Press, 1969); Immanuel Wallerstein, *The Modern World System*, vol. 1 (New York: Academic, 1974).

It should also be noted that the Soviet historian of science Boris Hessen had been making claims back in 1932 about the indirect positive effects on the growth of modern science in Europe from the emergence within Europe of protocapitalism's patterns of mercantile expansion—the need for better roads, better navigation, money, better firearms, and so on. See Boris Hessen, *The Economic Roots of Newton's Principia* (New York: Howard Fertig, 1970). Hessen's argument, apparently constituted within classical Marxist views of relations between capitalism and "bourgeois science," has conventionally been perceived as a prime example of an "externalist" one. However, one can wonder if today it could reasonably be reread as unintentionally heretical even to classical Marxism in proposing not external "influences" on a completely malleable science, but, rather, as much more the kind of "co-evolving science and society" argument characteristic of post-Kuhnian studies.

9. See chapter 1 of Lucille Brockway, *Science and Colonial Expansion: The Role of the British Royal Botanical Gardens* (New York: Academic, 1979); Roy MacLeod, "On Visiting the 'Moving Metropolis': Reflections on the Architecture of Imperial Science," in *Scientific Colonialism: A Cross Cultural Comparison,* ed. Nathan Reingold and Marc Rothenberg (Washington, D.C.: Smithsonian Institution Press, 1987).

10. See, e.g., Susantha Goonatilake, *Aborted Discovery: Science and Creativity in the Third World* (London: Zed, 1984); George Gheverghese Joseph, *The Crest of the Peacock: Non-European Roots of Mathematics* (New York: I. B. Tauris, 1991); Joseph Needham et al., *Science and Civilisation in China* (Cambridge: Cambridge University Press, 1954); Joseph Needham, *The Grand Titration: Science and Society in East and West* (Toronto: University of Toronto Press, 1969).

11. For examples from post-Kuhnian studies, see the papers in *Disunity,* ed. Galison and Stump.

12. See, e.g., Goonatilake, *Aborted;* Wendy Harcourt, ed., *Feminist Perspectives on Sustainable Development* (London: Zed, 1994); Maria Mies, *Patriarchy and Accumulation on a World Scale: Women in the International Division of Labor* (Atlantic Highlands, N.J.: Zed, 1986); Ashis Nandy, ed., *Science, Hegemony, and Violence: A Requiem for Modernity* (Delhi: Oxford, 1990); Wolfgang Sachs, ed., *The Development Dictionary: A Guide to Knowledge as Power* (Atlantic Highlands, N.J.: Zed, 1992); Ziauddin Sardar, ed., *The Revenge of Athena: Science Exploitation, and the Third World* (London: Mansell, 1988).

13. Cf., e.g., Blaut, *Colonizer's Model.* Blaut uses the phrases "multi-stream" and "single-stream" to characterize these eurocentric and anti-eurocentric kinds of histories. However, for readers who have in mind that "multicultural" is anti-eurocentric and "monocultural" is eurocentric, the references of Blaut's phrases can be hard to keep in mind. Hence, I have substituted "isolationist" versus "interactional" for his terms.

14. See Dorothy Smith, *The Conceptual Practices of Power: A Feminist Sociology of Knowledge* (Boston: Northeastern University Press, 1990) for one important discussion of how knowledge projects are always organized *from* some determinate social location which invariably leaves its fingerprints on the patterns and forms of systematic knowledge and ignorance produced. Samir Amin's *Eurocentrism* (New York: Monthly Review Press, 1989), and Edward Said's *Orientalism* (New York: Pantheon, 1978), are well-known discussions of how western imperial projects have organized distinctive patterns of knowledge and ignorance, their institutions, cultures, and practices.

15. This is not the place to go into why such divergent patterns of engagement exist. However, one can briefly note differences in the institutional origins of the two kinds of studies. The postcolonial accounts mostly have not been emerging from within European and American universities. Nor have many postcolonial accounts emerged from participant-observer ethnographies of modern science, where permission of the scientists is required for continued research.

16. I follow here David Hess's example in *Science and Technology in a Multicultural World: The Cultural Politics of Facts and Artifacts* (New York: Columbia University Press, 1995), 1.

17. William Whewell, *The Philosophy of the Inductive Sciences, Founded upon Their History* (London: Parker, 1840).

18. Or does this position actually depend upon the restricted notion of "science" criticized above?

19. Cf., e.g, David Wellman's classic account of empirical and theoretical problems with conceptualizing racism as merely prejudices in "Prejudiced People Are Not the Only Racists in America," the introduction to his *Portraits of White Racism* (New York: Cambridge University Press, 1977). Similar accounts can be found in Amin, *Eurocentrism*, Blaut, *Colonizer's Mind*, and Said, *Orientalism*.

20. A classic account is Ernest Gellner's "Concepts and Society" of how the concept of nobility functions among the Nuer (and, implicitly, in any aristocracy), in *Rationality*, ed. Bryan Wilson (New York: Harper and Row, 1970).

21. See James Joseph Scheurich and Michelle D. Young, "Coloring Epistemologies: Are Our Research Epistemologies Racially Biased?" *Educational Researcher* 26:3 (1997), 4–16. This analysis is valuable for its clarity about the differences between these five forms of "bias," though it lacks a complex analysis, e.g., of the kind to be found in Wellman 1977, of why it is that racist or otherwise biased explanations of social inequality are so widespread among peoples who also so value democratic ethics and politics.

22. One place where a balanced discussion of such poststructuralist conceptions of discourse can be found that is accessible to readers new to this field is in Chris Weedon's account of what feminist theory does and does not need from poststructuralism in her *Feminist Practice and Poststructuralist Theory* (Cambridge, Mass.: Blackwell, 1987).

23. See Amin, *Eurocentrism*, Blaut, *Colonizer's Mind*, and Said, *Orientalism*, for accounts of such material effects of civilizational eurocentrism.

24. Note Stephen Jay Gould's similar analysis in his *The Mismeasure of Man* (New York: Norton, 1981) of how good attitudes and state-of-the-art (for their day) scientific reputations did not protect nineteenth-century scientists against thinking craniology to be good science.

25. One familiar classic in feminist circles is Gloria Anzaldua's *Borderlands/La Frontera* (San Francisco: Spinsters/Aunt Lute, 1987).

26. The possibilities listed in the preceding two paragraphs are examined in *Colonial Discourse and Post-Colonial Theory*, ed. Patrick Williams and Laura Chrisman (New York: Columbia University Press, 1994). See especially the essays by Misha and Hodge, and McClintock. See also *The Post-Colonial Studies Reader*, ed. Bill Ashcroft, Gareth Griffiths, and Helen Tiffin (New York: Routledge, 1995).

27. Hess, *Science*.

28. *Gender, feminist*, and related terms will be discussed in subsequent chapters where the northern and southern feminist science and technology studies are explored.

29. See Anzaldua, *Borderlands;* Patricia Hill Collins, *Black Feminist Thought: Knowledge, Consciousness, and the Politics of Empowerment* (New York: Routledge, 1991).

30. Standpoint theory will be discussed more fully in a later chapter. Important recent articulations of it as such can be found in Collins, *Black Feminist;* Nancy Hartsock, "The Feminist Standpoint," in *Discovering Reality: Feminist*

Perspectives on Epistemology, Metaphysics, Methodology, and Philosophy of Science (Dordrecht: Reidel/Kluwer, 1983); Hilary Rose, "Hand, Brain, and Heart: A Feminist Epistemology for the Natural Sciences," *Signs* 9:1 (1983); and Dorothy Smith, *Conceptual Practices,* and *The Everyday World as Problematic: A Sociology for Women* (Boston: Northeastern University Press, 1987).

31. For example, the historically influential writings of John Stuart Mill and Marx and Engels routinely are taught in introduction to feminist theory courses.

32. This thorny issue will be pursued in a later chapter. Such arguments have been made by many science observers and knowledge theorists. See, for example, Richard Bernstein, *Beyond Objectivism and Relativism* (Philadelphia: University of Pennsylvania Press, 1983), and Sandra Harding, "After the Neutrality Ideal: Science, Politics and 'Strong Objectivity,'" *Social Research* 59 (1992), 567–87.

33. See, e.g., Robert Proctor, *Cancer Wars: How Politics Shapes What We Know and Don't Know about Cancer* (Boston: Basic, 1995).

34. See, e.g., *Disunity,* ed. Galison and Stump.

2. Postcolonial Science and Technology Studies

1. This chapter and the next draw on a number of my earlier papers, especially "Is Science Multicultural?" *Configurations* 2:2 (1994), and in *Multiculturalism: A Critical Reader,* ed. David Theo Goldberg (Oxford: Blackwell, 1994); "Is Modern Science an Ethnoscience?" in *Science and Technology in a Developing World,* ed. T. Shinn et al. (Dordrecht: Kluwer, 1997), and *Postcolonial African Philosophy,* ed. Emmanuel Chukwudi Eze (Oxford: Blackwell, 1996); and "European Expansion and the Organization of Modern Science in Europe," *Organization* special issue on Race, Gender, and Organization, ed. Marta Calas and Linda Smirchich, 3:4 (1996).

2. J. M. Blaut describes "the tunnel of time" in *The Colonizer's Model of the World: Geographical Diffusionism and Eurocentric History* (New York: Guilford Press, 1993). See also Eric Wolf, *Europe and the People without History* (Berkeley: University of California Press, 1984).

3. Blaut, *The Colonizer's Model.* See also J. M. Blaut, with contributions by Andre Gunder Frank, Samir Amin, Robert A. Dodgshon, and Ronen Palan, *1492: The Debate on Colonialism, Eurocentrism, and History* (Trenton: Africa World Press, 1992).

4. See Gary Nash et al., *National Standards for History* (Los Angeles: National Center for History in the Schools, 1996).

5. Thomas S. Kuhn, *The Structure of Scientific Revolutions,* 2d ed. (Chicago: University of Chicago Press, 1970).

6. The next chapter looks in more detail at new accounts of the relation between the successes of European expansion and of early modern sciences and technologies.

7. See, for example, some of these disputes in the contributions by other authors to Blaut's *1492.*

8. See the more extended discussion of "the postcolonial" in chapter 1 and in, e.g., Patrick Williams and Laura Chrisman, eds., *Colonial Discourse and Post-Colonial Theory* (New York: Columbia University Press, 1994).

9. See, e.g., Blaut, *The Colonizer's Model,* and the diverse contributions to Blaut, *1492.* Other resources for the critique of diffusionism can be found in Michael Adas, *Machines As the Measure of Man* (Ithaca: Cornell University Press, 1989); Morris Berman, *The Reenchantment of the World* (Ithaca: Cornell University

Press, 1981); Lucille H. Brockway, *Science and Colonial Expansion: The Role of the British Royal Botanical Gardens* (New York: Academic, 1979); Susantha Goonatilake, *Aborted Discovery: Science and Creativity in the Third World* (London: Zed, 1984), and "The Voyages of Discovery and the Loss and Rediscovery of the 'Other's Knowledge,'" *Impact of Science on Society*, no. 167 (1992), 241–64, and his *Toward a Global Science* (Bloomington: Indiana University Press, 1998); David J. Hess, *Science and Technology in a Multicultural World: The Cultural Politics of Facts and Artifacts* (New York: Columbia University Press, 1995); George Gheverghese Joseph, *The Crest of the Peacock: Non-European Roots of Mathematics* (New York: I. B. Tauris, 1991); R. K. Kochhar, "Science in British India," parts I and II, *Current Science* 63:11 (1992), 689–94; 64:1 (1993), 55–62; Deepak Kumar, *Science and Empire: Essays in Indian Context 1700–1947* (Delhi, India: Anamika Prakashan, and National Institute of Science, Technology, and Development, 1991); Donald F. Lach, *Asia in the Making of Europe*, vol. 2 (Chicago: University of Chicago Press, 1977); James E. McClellan, *Colonialism and Science: Saint Domingue in the Old Regime* (Baltimore: Johns Hopkins University Press, 1992); Ashis Nandy, ed., *Science, Hegemony, and Violence: A Requiem for Modernity* (Delhi: Oxford, 1990); Joseph Needham, *The Grand Titration: Science and Society in East and West* (Toronto: University of Toronto Press, 1969), and his *Science and Civilisation in China*, vols. 1–7 (Cambridge: Cambridge University Press, 1954–); Patrick Petitjean et al., eds., *Science and Empires: Historical Studies about Scientific Development and European Expansion* (Dordrecht: Kluwer, 1992); Nathan Reingold and Marc Rothenberg, eds., *Scientific Colonialism: A Cross-Cultural Comparison* (Washington, D.C.: Smithsonian Institution Press, 1987); Sal Restivo, *Mathematics in Society and History: Sociological Inquiries* (Dordrecht: Kluwer, 1992); Colin A. Ronan, *Lost Discoveries: The Forgotten Sciences of the Ancient World* (London: MacDonald, 1973); I. A. Sabra, "The Scientific Enterprise," in *The World of Islam*, ed. B. Lewis (London: Thames and Hudson, 1976); Ziauddin Sardar, ed., *The Revenge of Athena: Science, Exploitation, and the Third World* (London: Mansell, 1988); Immanuel Wallerstein, *The Modern World System: Capitalist Agriculture and the Origins of the European World Economy in the Sixteenth Century* (New York: Academic, 1974); Jack McIver Weatherford, *Indian Givers: What the Native Americans Gave to the World* (New York: Crown, 1988); Eric Wolf, *Europe and the People without History* (Berkeley: University of California Press, 1984); Frances Yates, *Giordano Bruno and the Hermetic Tradition* (New York: Vintage, 1969).

10. Blaut, *Colonizer's Model*, 50.

11. Ibid.

12. E. G. Lucien Levy-Bruhl, *Les Fonctions Mentales dans les sociétés inferieures* (Paris: Presses Universitaires de France, 1910), trans. as *How Natives Think* (London: Allen and Unwin, 1926).

13. Robin Horton, "African Traditional Thought and Western Science," pts. 1 and 2, *Africa* 37 (1967). But see also J. E. Wiredu, "How Not to Compare African Thought with Western Thought," in *African Philosophy*, 2d ed., ed. Richard A. Wright (Washington, D.C.: University Press of America, 1979).

14. See, e.g., Yates, *Giordano Bruno*.

15. Boris Hessen, *The Economic Roots of Newton's Principia* (New York: Howard Fertig, 1970).

16. Edgar Zilsel, "The Sociological Roots of Science," *American Journal of Sociology* 47 (1942).

17. This is recounted in Steven Shapin, *A Social History of Truth* (Chicago: University of Chicago Press, 1994).

18. Zilsel, "Sociological Roots"; Werner Van den Daele, "The Social Construc-

tion of Science," in *The Social Production of Scientific Knowledge*, ed. E. Mendelsohn, P. Weingart, and R. Whitley (Dordrecht: Reidel, 1977).

19. Needham, *Grand Titration*.

20. Francesca Bray, "Eloge for Joseph Needham," *Isis* 87:2 (1996), 312–17.

21. Mention should also be made of the work of the socialist scientist and science theorist, John Desmond Bernal. His *The Social Functions of Science* (London: Routledge and Kegan Paul, 1939) examined the inevitable contradiction Marxists perceived between modern science's productive forces and a capitalist political economy. Science could not be developed to its full creative potential within a capitalist order, they argued.

22. Blaut, *Colonizer's Model*, 203.

23. Ibid.

24. Ibid. (quotation). C. L. R. James, *The Black Jacobins: Toussaint L'Ouverture and the San Domingo Revolution*, 2d ed. rev. (New York: Vintage, 1963); Eric Williams, *Capitalism and Slavery* (Chapel Hill: University of North Carolina Press, 1994).

25. George Basalla, "The Spread of Western Science," *Science* 156 (May 1967), 611–22; see Roy MacLeod, "On Visiting the 'Moving Metropolis': Reflections on the Architecture of Imperial Science," in *Scientific Colonialism*, ed. Reingold and Rothenberg.

26. Frank, *Capitalism and Underdevelopment*, 4, quoted by Brockway, *Science*, 14–15.

27. Blaut, *Colonizer's Model*, 206.

28. Such an approach can be found, e.g., in C. Frake, "The Ethnographic Study of Cognitive Systems," in *Anthropology and Human Behaviour*, ed. T. Gladwin (Washington, D.C.: Anthropology Society of Washington, 1962). One example of recent illuminating uses of such comparative approaches can be found in the work of David Turnbull and Helen Watson-Verran. See David Turnbull, "Local Knowledge and Comparative Scientific Traditions," *Knowledge and Policy* 6:3/4 (1993), 29–54, and Helen Watson-Verran and David Turnbull, "Science and Other Indigenous Knowledge Systems," in *Handbook of Science and Technology Studies*, ed. S. Jasanoff, G. Markle, T. Pinch, and J. Petersen (Thousand Oaks, Calif.: Sage, 1995), 115–39. This last essay also provides an account of the differences between this later work and the earlier comparative ethnoscience approaches.

29. The following lists appeared in earlier forms in my "Is Science Multicultural?"

30. Yates, *Giordano Bruno*.

31. Lach, *Asia in the Making*; Seyyed Hossein Nasr, "Islamic Science, Western Science: Common Heritage, Diverse Destinies," in Sardar, ed., *Revenge*; Sabra, "The Scientific Enterprise."

32. See Weatherford, *Indian Givers*.

33. See Adas, *Machines As the Measure*; Joseph, *Crest*.

34. Goonatilake, *Aborted Discovery*; Needham, *Grand Titration*, and *Science and Civilisation*.

35. Kochhar, "Science," vol. 1, 694. Additional long and detailed lists of the borrowings pointed to in the above paragraphs can be found in the sources cited here and earlier.

36. And, as V. Y. Mudimbe pointed out to me, of Europe itself, for European sciences also constituted European lands, cities, and peoples as their laboratories. Consider, for example, the way women, the poor, children, the sexually "deviant," the sick, the mad, rural and urban populations, and workers have continuously generated information under the observing eyes of sciences.

37. Goonatilake, "Voyages."

38. See, e.g., Harcourt, *Feminist Perspectives;* Sachs, *Development Dictionary;* and many of the essays in Sardar, ed., *Revenge.*

39. See especially chapters 5 and 7.

3. Voyages of Discovery

1. James E. McClellan, *Colonialism and Science: Saint Domingue in the Old Regime* (Baltimore: Johns Hopkins University Press, 1992), 1, 7.

2. Alfred Crosby, *Ecological Imperialism: The Biological Expansion of Europe, 900–1900* (Cambridge: Cambridge University Press, 1987). See also his *The Columbian Exchange: Biological and Cultural Consequences of 1492* (Westport: Greenwood, 1972).

3. A nearby museum contains some of the five-century-old "letter stones" of the first Portuguese and other Europeans to round the Cape—stones left as messages for subsequent voyagers, on which are carved the name of the ship and the date on which the voyagers set foot on this coast.

4. J. M. Blaut, *The Colonizer's Model of the World: Geographical Diffusionism and Eurocentric History* (New York: Guilford Press, 1993), 153–78. Crosby's account supports Blaut's. See also Blaut, *1492: The Debate on Colonialism, Eurocentrism, and History* (with contributions by Andre Gunder Frank, Samir Amin, Robert A. Dodgshon, Ronen Palan, and Peter Taylor) (Trenton, N.J.: Africa World Press, 1992).

5. Blaut, *Colonizer's Model,* 184.

6. Ibid., 15.

7. Lucille H. Brockway, *Science and Colonial Expansion: The Role of the British Royal Botanical Gardens* (New York: Academic, 1979).

8. McClellan, *Colonialism and Science.*

9. R. K. Kochhar, "Science in British India" *Current Science,* part I, 63:11, (1992), 689–94; part II, 64:1, (1993), 55–62.

10. McClellan, *Colonialism,* 289.

11. Kochhar, "Science," part II, 61.

12. Brockway, *Science,* 127.

13. Ibid.

14. Kochhar, "Science," part I, 689.

15. Ibid.

16. McClellan, *Colonialism,* 290.

17. Kochhar, "Science," part I, 694.

18. Brockway, *Science,* 18, quoting Carlo M. Cipollo, *Before the Industrial Revolution: European Society and Economy, 1000–1700* (New York: Norton, 1976), 214.

19. Ibid.

20. Brockway, *Science,* 36. See also Crosby, *Columbian Exchange.*

21. Brockway, *Science,* 36.

22. Ibid.

23. Ibid., 8.

24. The claims of this and the next paragraph can be found, for example, in Deepak Kumar, *Science and Empire: Essays in Indian Context, 1700–1947* (Delhi, India: Anamika Prakashan, and National Institute of Science, Technology, and Development, 1991); Susantha Goonatilake, *Aborted Discovery: Science and Creativity in the Third World* (London: Zed, 1984); Ashis Nandy, ed., *Science, Hegemony, and Violence: A Requiem for Modernity* (Delhi: Oxford, 1990); Patrick Petitjean et al., eds., *Science and Empires: Historical Studies about Scientific Development and European Expansion* (Dordrecht: Kluwer, 1992); Walter Rodney, *How*

Europe Underdeveloped Africa (Washington, D.C.: Howard University Press, 1982); Ziauddin Sardar, ed., *The Revenge of Athena: Science, Exploitation, and the Third World* (London: Mansell, 1988); Immanuel Wallerstein, *The Modern World System*, vol. 1 (New York: Academic, 1974).

25. See Crosby, *Ecological Imperialism* for these cases.

26. Brockway, *Science*, 195.

27. Thomas S. Kuhn, *The Structure of Scientific Revolutions*, 2d ed. (Chicago: University of Chicago Press, 1970).

28. See Eric Wolf, *Europe and the People without History* (Berkeley: University of California Press, 1984) on a similar common attitude toward the European peasantry.

29. For examples of some of the well-known post-Kuhnian accounts that converge with the postcolonial studies in raising the following philosophic issues, see Paul Feyerabend, *Against Method* (London: New Left, 1975); Peter Galison, *How Experiments End* (Chicago: University of Chicago Press, 1987); *The Disunity of Science*, ed. Peter Galison and David J. Stump (Stanford: Stanford University Press, 1996); Ian Hacking, *Representing and Intervening* (Cambridge: Cambridge University Press, 1983); Donna Haraway, *Primate Visions: Gender, Race, and Nature in the World of Modern Science* (New York: Routledge, 1989); N. Katherine Hayles, "Constrained Constructivism: Locating Scientific Inquiry in the Theater of Representation," in *Realism and Representation*, ed. George Levine (Madison: University of Wisconsin Press, 1993); Mary Hesse, *Models and Analogies in Science* (Notre Dame: University of Notre Dame Press, 1966); Evelyn Fox Keller, *Reflections on Gender and Science* (New Haven: Yale University Press, 1984); Bruno Latour, *Science in Action* (Cambridge: Harvard University Press, 1987), and *The Pasteurization of France* (Cambridge: Harvard University Press, 1988); Bruno Latour and Steve Woolgar, *Laboratory Life: The Social Construction of Scientific Facts* (Beverly Hills: Sage, 1979); Andrew Pickering, ed., *Science as Practice and Culture* (Chicago: University of Chicago Press, 1992), and his *Constructing Quarks* (Chicago: University of Chicago Press, 1984); Robert Proctor, *Value-Free Science? Purity and Power in Modern Knowledge* (Cambridge: Harvard University Press, 1991); Sal Restivo, *Mathematics in Society and History* (Dordrecht: Kluwer, 1992); Joseph Rouse, *Knowledge and Power: Toward a Political Philosophy of Science* (Ithaca: Cornell University Press, 1987); John A. Schuster and Richard R. Yeo, eds., *The Politics and Rhetoric of Scientific Method: Historical Studies* (Dordrecht: Reidel, 1986); Steven Shapin, *A Social History of Truth* (Chicago: University of Chicago Press, 1994); Sharon Traweek, *Beamtimes and Lifetimes* (Cambridge, Mass.: MIT Press, 1988), and her "An Introduction to Cultural, Gender, and Social Studies of Sciences and Technologies," in *Culture, Medicine, and Psychiatry* 17 (1993), 3–25; Steve Woolgar, *Science: The Very Idea* (New York: Tavistock, 1988).

4. Cultures as Toolboxes for Sciences and Technologies

1. One should recollect here that "southern"(or "postcolonial") and "northern" in this account refer to *schools* of science studies, to discursive traditions, not to the ethnicity, nationality, or residence of their practitioners, which are diverse.

2. Edward Said, *Orientalism* (New York: Pantheon, 1978). Susantha Goonatilake discusses the value of identifying the culturally distinctive characteristics of European (northern) sciences and technologies in a kind of mirror image of the "orientologies" that Europeans have produced in his *Aborted Discovery:*

Science and Creativity in the Third World (London: Zed, 1984). However, in contrast to orientologies, Europologies are not intended to caricature or devalue European culture, but merely to identify the elements of European sciences and technologies that are indeed local rather than, as conventional eurocentric assumptions would have it, universal.

3. This section is a lightly edited version of a passage from my "Is Science Multicultural? Challenges, Resources, Opportunities, Uncertainties," in *Configurations* 2:2 (1994), and in *Multiculturalism: A Critical Reader,* ed. David Theo Goldberg (Oxford: Blackwell, 1994).

4. Joseph Needham, "The Laws of Nature and the Laws of Man," in *The Grand Titration: Science and Society in East and West* (Toronto: University of Toronto Press, 1969), 302.

5. Does it also reflect the modern familiarity with large bureaucracies and corporations where individual responsibility for rules and regulations is often unlocatable? That is, would the idea of characterizing nature's order merely through statistical regularities have been comprehensible or attractive to people living in absolute monarchies? Would it have been perceived to be politically threatening to such monarchies?

6. Evelyn Fox Keller, *Reflections on Gender and Science* (New Haven: Yale University Press, 1984), 131, 132.

7. R. K. Kochhar, "Science in British India," part I, *Current Science* 63:11 (1992), 694.

8. See, e.g., Paul Forman, "Behind Quantum Electronics: National Security as Bases for Physical Research in the U.S., 1940–1960," *Historical Studies in Physical and Biological Sciences* 18 (1987). Cf. also Michael Adas, *Machines As the Measure of Man* (Ithaca: Cornell University Press, 1989); Samir Amin, *Eurocentrism* (New York: Monthly Review Press, 1989); Lucille H. Brockway, *Science and Colonial Expansion: The Role of the British Royal Botanical Gardens* (New York: Academic, 1979); Alfred Crosby, *Ecological Imperialism: The Biological Expansion of Europe* (Cambridge: Cambridge University Press, 1987); Goonatilake, *Aborted Discovery,* and "The Voyages of Discovery and the Loss and Rediscovery of the 'Other's' Knowledge," *Impact of Science on Society* no. 167 (1992), 241–64; Daniel R. Headrick, ed., *The Tools of Empire: Technology and European Imperialism in the Nineteenth Century* (New York: Oxford University Press, 1981); James E. McClellan, *Colonialism and Science: Saint Domingue in the Old Regime* (Baltimore: Johns Hopkins University Press, 1992); Ashis Nandy, ed., *Science, Hegemony, and Violence: A Requiem for Modernity* (Delhi: Oxford University Press, 1990); Patrick Petitjean et al., eds., *Science and Empires: Historical Studies about Scientific Development and European Expansion* (Dordrecht: Kluwer, 1992); Ziauddin Sardar, ed., *The Revenge of Athena: Science, Exploitation, and the Third World* (London: Mansell, 1988).

9. Brockway, *Science.*

10. See, e.g., Rustum Roy's discussions about what should be considered to constitute "fundamental physics" generated in the debate over funding for the proposed superconducting super collider in his untitled letters, *Physics Today* 38:9 (1985), 9–11; 39:2 (1986), 15, 96; 39:4 (1986), 81–84; 40:2 (1987), 13.

11. Kok Peng Khor, "Science and Development: Underdeveloping the Third World," in *Revenge,* ed. Sardar, 207–208.

12. Claude Alvares, "Science, Colonialism, and Violence: A Luddite View," in *Science,* ed. Nandy, 108.

13. J. Bandyopadhyay and V. Shiva, "Science and Control: Natural Resources and Their Exploitation," in *Revenge,* ed. Sardar, 63.

14. Nancy Hartsock, "The Feminist Standpoint," in *Discovering Reality*, ed. Sandra Harding and Merrill Hintikka (Dordrecht: Reidel/Kluwer, 1983); Dorothy Smith, *The Conceptual Practices of Power: A Feminist Sociology of Knowledge* (Boston: Northeastern University Press, 1990); Alfred Sohn-Rethel, *Intellectual and Manual Labor* (London: Macmillan, 1978).

15. Bandyopadhyay and Shiva, "Science," 60.

16. So-called biological differences in humans are themselves the product also of social relations. See, e.g., Helen Lambert, "Biology and Equality: A Perspective on Sex Differences," in *Sex and Scientific Inquiry*, ed. Sandra Harding and Jean O'Barr (Chicago: University of Chicago Press, 1987), and section 2 of Sandra Harding, ed., *The "Racial" Economy of Science: Toward a Democratic Future* (Bloomington: Indiana University Press, 1993).

17. Crosby, *Ecological Imperialism*.

18. J. M. Blaut, *The Colonizer's Model of the World: Geographical Diffusionism and Eurocentric History* (New York: Guilford Press, 1993); cf. Crosby, *Ecological Imperialism*.

19. "Same" is in scare quotes since, as we shall see in the next two sections also, different cultures conceptualize "the world" in ways that are shaped by their distinctive interests, discursive resources, and ways of organizing the production of knowledge about nature. In important senses, different interests, discursive resources, and ways of organizing knowledge production create different worlds, different patterns "in nature" that are the object of systematic scrutiny. Many such patterns are consistent with nature's order, but none are uniquely congruent with it—as we shall shortly see.

20. Donna Haraway, *Primate Visions: Gender, Race, and Nature in the World of Modern Science* (New York: Routledge, 1989).

21. See Robert Proctor, *Cancer Wars: How Politics Shapes What We Know and Don't Know about Cancer* (Boston: Basic, 1995); Joni Seager, *Earth Follies: Coming to Feminist Terms with the Global Environmental Crisis* (New York: Routledge, 1993).

22. Forman, "Quantum Electronics."

23. Evelyn Fox Keller, "Gender and Science," in *Discovering Reality*, ed. Harding and Hintikka, 235.

24. National Academy of Sciences, *On Being a Scientist* (Washington, D.C.: National Academy Press, 1989), 5–6.

25. John Dupre, *The Disorder of Things: Metaphysical Foundations for the Disunity of Science* (Cambridge: Harvard University Press, 1993), 6.

26. Other topics of such analyses include genres of science and technology writings and the rhetorics of sciences and technologies.

27. See, for example, Paul Feyerabend, *Against Method* (London: New Left, 1975); Haraway, *Primate Visions*; Sandra Harding, *The Science Question in Feminism* (Ithaca: Cornell University Press, 1986); Keller, *Reflections*; Thomas Kuhn, *The Structure of Scientific Revolutions*, 2d ed. (Chicago: University of Chicago Press, 1970); Bruno Latour, *The Pasteurization of France* (Cambridge: Harvard University Press, 1988); Emily Martin, *The Woman in the Body: A Cultural Analysis of Reproduction* (Boston: Beacon, 1987); Carolyn Merchant, *The Death of Nature: Women, Ecology, and the Scientific Revolution* (New York: Harper and Row, 1980); Andrew Pickering, ed., *Science as Practice and Culture* (Chicago: University of Chicago Press, 1992); Steven Shapin and Simon Schaffer, *Leviathan and the Air Pump* (Princeton: Princeton University Press, 1985).

28. Mary Hesse, *Models and Analogies in Science* (Notre Dame: University of Notre Dame Press, 1966); Nancy Leys Stepan, "Race and Gender: The Role of Analogy in Science," *Isis* 77 (1986); Bas Van Fraassen and Jill Sigman, "Interpre-

tation in Science and in the Arts," in *Realism and Representation*, ed. George Levine (Madison: University of Wisconsin Press, 1993).

29. Haraway, *Primate Visions*.

30. Michael Baxandall, *Painting and Experience in Fifteenth-Century Italy*, 2d ed. (Oxford: Oxford University Press, 1988), 45–56, 81–108.

31. Shapin and Schaffer, *Leviathan*.

32. Latour, *Pasteurization*.

33. Sharon Traweek, *Beamtimes and Lifetimes* (Cambridge, Mass.: MIT Press, 1988).

34. Kuhn, *Structure*.

35. Should these positions of mere cultural difference be counted as standpoints along with the positions of the exploited and marginalized that traditionally have been the object of standpoint analyses (to be pursued in a later chapter)? They share being outside the prevailing conceptual frameworks, and requiring struggles to grasp, but they can differ in the degree to which they are stratified by power relations. Standpoint theory importantly is a theory about the relations between power and knowledge, not just difference and knowledge. I leave the question unanswered!

5. Postcolonial Feminist Science Studies

1. This chapter draws on the first and last sections especially of Sandra Harding, "Multicultural and Global Feminist Philosophies of Science: Resources and Challenges," in *Feminism, Science, and the Philosophy of Science*, ed. Lynn Hankinson Nelson and Jack Nelson (Dordrecht: Kluwer, 1996).

2. For the classic four public agenda feminist theories, see Alison Jaggar, *Feminist Politics and Human Nature* (Totowa, N.J.: Rowman and Allenheld, 1983). The most recent edition of Alison Jaggar and Paula Rothenberg, *Feminist Frameworks* (New York: McGraw Hill, 1993), includes brief excerpts from classic writings of all six of these feminist approaches. Just a few examples of influential writings in multicultural and global feminisms from a variety of disciplines are: Gloria Anzaldua, *Borderlands/La Frontera* (San Francisco: Spinsters/Aunt Lute, 1987); Anzaldua's edited *Making Face/Making Soul* (San Francisco: Aunt Lute Foundation Press, 1990); Patricia Hill Collins, *Black Feminist Thought: Knowledge, Consciousness, and the Politics of Empowerment* (New York: Routledge, 1991); Cynthia Enloe, *Bananas, Beaches, and Bases: Making Feminist Sense of International Politics* (Berkeley: University of California Press, 1990); Donna Haraway, *Primate Visions: Gender, Race, and Nature in the World of Modern Science* (New York: Routledge, 1989); Wendy Harcourt, ed., *Feminist Perspectives on Sustainable Development* (London: Zed, 1994); bell hooks, *Ain't I a Woman? Black Women and Feminism* (Boston: South End Press, 1981); Maria Mies, *Patriarchy and Accumulation on a World Scale: Women in the International Division of Labor* (Atlantic Highlands, N.J.: Zed, 1986); Gita Sen and Caren Grown, *Development Crises and Alternative Visions: Third World Women's Perspectives* (New York: Monthly Review Press, 1987); Vandana Shiva, *Staying Alive: Women, Ecology, and Development* (London: Zed, 1989); Pamela Sparr, ed., *Mortgaging Women's Lives: Feminist Critiques of Structural Adjustment* (London: Zed, 1994). Of course there are many "postcolonialisms," as noted in an earlier chapter. Good dialogues on this issue can be found in Patrick Williams and Laura Chrisman, eds., *Colonial Discourse and Post-Colonial Theory* (New York: Columbia University Press, 1994), and *The Post-Colonial Studies Reader*, ed. Bill Ashcroft, Gareth Griffith, and Helen Tiffin (New York: Routledge, 1995). Here the term refers to a space that has been

opened up by historical changes and the discussions created by them—a space in which new issues and new questions can be raised.

3. I first distinguished what I called feminist empiricism from other feminist epistemologies emerging at the time in chapter 2 of *The Science Question in Feminism* (Ithaca: Cornell University Press, 1986). I had in mind the telling criticisms of the carelessness in applying rigorously existing standards for good method that were emerging from practicing biologists and social scientists during the 1970s and early 1980s. These critics did not give their epistemology a special name since they did not think of it as a distinctive epistemology. Subsequently, several philosophers developed powerful analyses of how feminist concerns about science and knowledge in general could be fit into the conceptual framework of empiricist philosophy with minimum stress to conventional empiricist assumptions. These were produced after such other feminist epistemologies as feminist standpoint theory and feminist poststructuralism had begun to be developed, and so were created to contrast with the other feminist theories of knowledge as much as with androcentric approaches. (It should be noted that all of these theories value *empirical* research. "Empiricism" refers to a five-century-old philosophy of science.) Among those who explicitly called their work *empiricist* were Lorraine Code, *What Can She Know?* (Ithaca: Cornell University Press, 1991); Helen Longino, *Science as Social Knowledge* (Princeton: Princeton University Press, 1990); and Lynn Hankinson Nelson, *Who Knows? From Quine to a Feminist Empiricism* (Philadelphia: Temple University Press, 1990). A number of collections of essays mainly elaborating and defending feminist empiricism have appeared; see, e.g., Louise Antony and Charlotte Witt, *A Mind of Her Own* (Boulder: Westview, 1993), and several British collections, e.g., Kathleen Lennon and Margaret Whitford, *Knowing the Difference: Feminist Perspectives on Epistemology* (New York: Routledge, 1994).

4. Important writings here include Nancy Hartsock, "The Feminist Standpoint," in *Discovering Reality,* ed. Sandra Harding and Merrill Hintikka (Dordrecht: Reidel/Kluwer, 1983); Alison Jaggar, *Feminist Politics and Human Nature* (Totowa, N.J.: Rowman and Allenheld, 1983), chapter 11; Hilary Rose, "Hand, Brain, and Heart: A Feminist Epistemology for the Natural Sciences," *Signs* 9:1 (1983); Dorothy Smith, *The Everyday World as Problematic: A Sociology for Women* (Boston: Northeastern University Press, 1987), and *The Conceptual Practices of Power* (Boston: Northeastern University Press, 1990). Smith's volumes collect papers that had been appearing since the early 1970s. Other writers who do not specifically refer to their work as a standpoint approach have also contributed to this account. Two notable examples here are Donna Haraway, especially in "Situated Knowledges" in *Simians, Cyborgs, and Women* (New York: Routledge, 1991), and Catharine MacKinnon, in "Feminism, Marxism, Method, and the State," *Signs* 7:3 (1982).

5. These three phrases were produced respectively by Collins in *Black Feminist Thought,* hooks in *Feminist Theory from Margin to Center,* and Anzaldua in *Borderlands.*

6. Note that these courses and discourses have not all been "on the left," since, for example, on many campuses business schools and international relations programs have their own reasons to have become major supporters of their own forms of multicultural and global ("international") studies. The political consequences of better understandings of multiculturalism and global political economies can be diverse.

7. Nancy Leys Stepan, "Race and Gender: The Role of Analogy in Science," *Isis* 77 (1986), 360–61; citations are to the reprint in Sandra Harding, *The "Racial"*

Economy of Science (Bloomington: Indiana University Press, 1993). Carl Vogt, *Lectures on Man: His Place in Creation, and in the History of the Earth* (London: Longman, Green and Roberts, 1864), 81.

8. Edward Said, *Orientalism* (New York: Pantheon, 1978).

9. Londa Schiebinger, *Nature's Body: Gender in the Making of Modern Science* (Boston: Beacon, 1993), especially chapters 3–6; Anne Fausto-Sterling, "Gender, Race, and Nation: The Comparative Anatomy of 'Hottentot' Women in Europe I: 1815–1817," in *Deviant Bodies*, ed. Jennifer Terry and Jacqueline Urla (Bloomington: Indiana University Press, 1995).

10. Donna Haraway, *Primate Visions: Gender, Race, and Nature in the World of Modern Science* (New York: Routledge, 1989); *Simians, Cyborgs, and Women: The Reinvention of Nature* (New York: Routledge, 1991); *Modest Witness@Second_Millennium.FemaleMan© Meets_OncoMouse™: Feminism and Technoscience* (New York: Routledge, 1997). The first two papers, "Animal Sociology and a Natural Economy of the Body Politic," parts I and II, reprinted in *Primate Visions*, originally appeared in *Signs* 4:1 and 4:2 (1978).

11. See Jack Stauder, "The 'Relevance' of Anthropology to Colonialism and Imperialism," in *Radical Science Essays*, ed. Les Levidow (London: Free Association, 1986); reprinted in *The "Racial" Economy*, ed. Harding.

12. Frederique Apffel-Marglin and Suzanne L. Simon, "Feminist Orientalism and Development," in *Feminist Perspectives*, ed. Harcourt, 32.

13. See, e.g., Enloe, *Bananas;* Harcourt, *Feminist Perspectives;* Mies, *Patriarchy;* Shiva, *Staying Alive;* Sparr, *Mortgaging.*

14. Examples of especially influential such studies are: Mary G. Belenky et al., *Women's Ways of Knowing: The Development of Self, Voice, and Mind,* 2d ed. (New York: Basic, 1997); Carol Gilligan, *In a Different Voice: Psychological Theory and Women's Development* (Cambridge: Harvard University Press, 1982); Catharine MacKinnon, "Feminism, Marxism, Method, and the State: An Agenda for Theory," *Signs* 7:3 (1982); Genevieve Lloyd, *The Man of Reason: "Male" and "Female" in Western Philosophy* (Minneapolis: University of Minnesota Press, 1984).

15. The European institutional context for feminist science and technology studies is somewhat different. Two causes are the higher discrimination against women in so many of the European universities and technical institutes and the higher state sponsorship of feminist research in at least some European countries, such as in Scandinavia. But there are no doubt further important institutional differences in the context for feminist science and technology research.

16. See, e.g., Harcourt, *Feminist Perspectives.*

17. See Anzaldua, *Borderlands;* Gloria Anzaldua and Cherrie Moraga, *This Bridge Called My Back* (New York: Women of Color Kitchen Table Press, 1981); Collins, *Black Feminist Thought;* Angela Davis, *Women, Race, and Class* (New York: Random House, 1983); hooks, *Feminist Theory;* Gloria Hull, Patricia Bell Scott, and Barbara Smith, *But Some of Us Are Brave: Black Women's Studies* (Old Westbury: Feminist Press, 1982); and Gayatri Spivak, *In Other Worlds: Essays in Cultural Politics* (New York: Methuen, 1987), for examples of influential texts developing multicultural and global conceptions of gender.

18. I include aspects of my own writings in the philosophy of science as the target of my comments here. Tom Patterson has suggested that "nature" is probably also a class concept, introduced to distinguish what educated classes of modern scientists know from what peasants and farmers know about the world around them, and to delegitimate the latter. This is a most interesting argument for the contrast noted here. See his "Nature: The Shadow of Civili-

zation" (forthcoming), and his *Inventing Western Civilization* (New York: Monthly Review Press, 1997).

19. Should we call these "mestiza sciences," in reference to Anzaldua's eloquent elaboration of the characteristics of peoples living on borderlands? See her *Borderlands/La Frontera*. It would be interesting to chart how many of the characteristics she delineates have an analog in characteristics of these "low" sciences and their practitioners.

6. Are There Gendered Standpoints on Nature?

1. This chapter appeared in an earlier form in *Osiris* 12 (1997).

2. Apparently even the distribution of "biological" sex differences in populations and the definition of what counts as a sex difference are highly shaped by gendered ways of organizing social relations and giving meaning to the world around us. See Judith Butler, *Gender Trouble: Feminism and the Subversion of Identity* (New York: Routledge, 1990); and Helen H. Lambert, "Biology and Equality: A Perspective on Sex Differences," *Signs* 4:1 (1978), reprinted in *Sex and Scientific Inquiry*, ed. Sandra Harding and Jean O'Barr (Chicago: University of Chicago Press, 1987).

3. See Sandra Harding, "Gender and Science: Two Problematic Concepts," in my *The Science Question in Feminism* (Ithaca: Cornell University Press, 1986).

4. I thank Barbara Laslett for pointing out to me that poverty can create forms of structural androgyny. See Maria Mies, *Patriarchy and Accumulation on a World Scale: Women in the International Division of Labor* (Atlantic Highlands, N.J.: Zed, 1986) on the recent "housewifization" of the (male) working class.

5. Of course there are biological constraints on all human activities and one cannot distinguish the biological from the social contributions to particular patterns of human belief and behavior. See, e.g., Lambert, "Biology and Equality." Moreover, patterns of behavior can leave their traces in the brain, as they do in other parts of the body. I shall not pursue here these well-known scientific facts.

6. Margaret Rossiter, *Women Scientists in America: Struggles and Strategies to 1940* (Baltimore: Johns Hopkins University Press, 1982).

7. For a discussion of the diverse reasons why sciences become de facto "women's sciences," see Ann Hibner Koblitz, "Challenges in Interpreting Data," in "The Gender Dimension of Science and Technology," ed. S. Harding and E. McGregor, a chapter in *UNESCO World Science Report 1996* (Paris: UNESCO, 1996).

8. A later chapter returns to discuss standpoint epistemology. See my "Rethinking Standpoint Epistemology," in *Feminist Epistemologies*, ed. L. Alcoff and E. Potter (New York: Routledge, 1992) for one place such issues have been addressed.

9. As noted in the earlier chapter, the four categories used here are selected because each has been the object of study of one or more widely recognized disciplinary focuses. Thus, we already know how to think about how these kinds of differences can shape different ranges of knowledge.

10. See, e.g., Emily K. Abel and Carole H. Browner, "Selective Compliance with Biomedical Authority and the Uses of Experiential Knowledge," in *Pragmatic Women and Body Politics*, ed. M. Lock and P. Kaufert (Cambridge: Cambridge University Press, 1998); Hilary Rose, "Hand, Brain, and Heart: A Feminist Epistemology for the Natural Sciences," *Signs* 9:1 (1983), and the feminist literature on women's reproductive technologies.

11. See Joni Seager, *Earth Follies: Coming to Feminist Terms with the Global Environmental Crisis* (New York: Routledge, 1993); Vandana Shiva, *Staying Alive: Women, Ecology, and Development* (London: Zed, 1989); Rosi Braidotti et al., *Women, the Environment, and Sustainable Development* (Atlantic Highlands, N.J.: Zed, 1994); Wendy Harourt, ed., *Feminist Perspectives on Sustainable Development* (London: Zed, 1994).

12. Evelyn Fox Keller, *A Feeling for the Organism* (San Francisco: Freeman, 1983).

13. To speak of "biological differences" here of course ignores the impossibility biologists point out of distinguishing biological from social contributions to human bodies. Note that that not all males are men, nor are all females women. Sex and gender categories are far from securely attached to each other in many cultures and subcultures as, for example, in the transgender cultures of transvestism, gender-"passing," and "gender-bending" in parts of contemporary U.S. and European culture, or in the generous varieties of gender recognized in various Native American groups.

14. See, e.g., N. Katherine Hayles, "Gender Encoding in Fluid Mechanics: Masculine Channels and Feminine Flows," *Differences* 4:2 (1992), 16–44; Evelyn Fox Keller, *Reflections on Gender and Science* (New Haven: Yale University Press, 1984), and her *Secrets of Life, Secrets of Death: Essays on Language, Gender, and Science* (New York: Routledge, 1992); Emily Martin, *The Woman in the Body: A Cultural Analysis of Reproduction* (Boston: Beacon, 1987); her *Flexible Bodies* (Boston: Beacon, 1994); and her "The Sperm and the Egg," *Signs* 16:3 (1991), 485–501; Carolyn Merchant, *The Death of Nature: Women, Ecology, and the Scientific Revolution* (New York: Harper and Row, 1980); Londa Schiebinger, *The Mind Has No Sex? Women in the Origins of Modern Science* (Cambridge: Harvard University Press, 1989), and her *Nature's Body: Gender in the Making of Modern Science* (Boston: Beacon, 1993); Sharon Traweek, *Beamtimes and Lifetimes* (Cambridge, Mass.: MIT Press, 1988).

15. See, for example, Brian Easlea, *Science and Sexual Oppression* (London: Weidenfeld and Nicolson, 1981), his *Witch-hunting, Magic and the New Philosophy* (Brighton, Sussex: Harvester Press, 1980), and his *Fathering the Unthinkable: Masculinity, Scientists and the Nuclear Arms Race* (London: Pluto Press, 1983); David Noble, *A World without Women: The Clerical Culture of Modern Science* (New York: Knopf, 1992), and his *The Religion of Technology* (New York: Knopf, 1995); Robert Nye, "Science and Medicine as Masculine Fields of Honor," *Osiris* 12 (1997).

16. Donna Haraway, *Primate Visions: Gender, Race, and Nature in the World of Modern Science* (New York: Routledge, 1989).

17. Sharon Traweek, *Big Science in Japan* (forthcoming).

18. Marcia Barinaga, "Is There a 'Female Style' in Science?" *Science* 260 (April 1993), 384–91.

19. See, e.g., J. Acker, "Gendered Institutions: From Sex Roles to Gendered Institutions," *Contemporary Sociology* 21 (1992) 565–69; S. Harding and E. McGregor, "The Gender Dimension of Science and Technology," *UNESCO World Science Report 1996* (Paris: UNESCO, 1996); R. M. Kanter, *Men and Women of the Corporation* (New York: Basic, 1977); A. J. Mills and P. Tancred, eds., *Gendering Organizational Theory* (London: Sage, 1991).

20. After all, ozone holes, the decrease in biodiversity, the use of nuclear weapons, and an increase in contagious diseases will have some different but mostly similar—where not identical—consequences for everyone, regardless of one's gender. It needs to be emphasized that the argument here is not that all

scientific projects are selected by gendered interests (though gendered discourses and ways of organizing research may still shape their findings).

21. See Rachel Carson, *Silent Spring* (Harmondsworth, U.K.: Penguin, 1962); Seager, *Earth Follies*.

7. Gender, Modernity, Knowledge

1. Joan Kelly-Gadol, "The Social Relation of the Sexes: Methodological Implications of Women's History," *Signs* 1:4 (1976), reprinted in *Feminism and Methodology*, ed. Sandra Harding (Bloomington: Indiana University Press, 1987), 16–17. See also Kelly-Gadol's "Did Women Have a Renaissance?" in *Becoming Visible*, ed. R. Bridenthal and C. Koonz (Boston: Houghton Mifflin, 1976).

2. For example, Emily K. Abel and Carole H. Browner, "Selective Compliance with Biomedical Authority and the Uses of Experiential Knowledge," in *Pragmatic Women and Body Politics*, ed. M. Lock and P. Kaufert (Cambridge: Cambridge University Press, 1998); Barbara Ehrenreich and Deirdre English, *Witches, Midwives and Nurses* (Old Westbury: Feminist Press, 1973); Ruth Hubbard, *The Politics of Women's Biology* (New Brunswick: Rutgers University Press, 1990); Hilary Rose, "Hand, Brain, and Heart: A Feminist Epistemology for the Natural Sciences," *Signs* 9:1 (1983).

3. Mary F. Belenky, B. M. Clinchy, N. R. Goldberger, and J. M. Tarule, *Women's Ways of Knowing*, 2d ed. (New York: Basic, 1986); Nancy Chodorow, *The Reproduction of Mothering* (Berkeley: University of California Press, 1978); Carol Gilligan, *In a Different Voice* (Cambridge: Harvard University Press, 1982); Nancy Goldberger et al., *Knowledge, Difference, and Power: Essays Inspired by Women's Ways of Knowing* (New York: Basic, 1996); Sara Ruddick, *Maternal Thinking* (Boston: Beacon, 1989).

4. Standpoint theories have proven especially useful in their concern to account for differences in patterns of knowledge and ignorance created by political relations—an issue to be explored further in a later chapter. The dominant groups in such political relations produce conceptual frameworks in public policy and research disciplines that value the local knowledge that their own activities and interests make reasonable to them, while devaluing and conceptually suppressing the patterns of knowledge and competing conceptual frameworks that emerge from the activities and interests of the groups disadvantaged by the power of the dominant groups. See Patricia Hill Collins, *Black Feminist Thought: Knowledge, Consciousness, and the Politics of Empowerment* (New York: Routledge, 1991); Sandra Harding, *The Science Question in Feminism* (Ithaca: Cornell University Press, 1986), and *Whose Science? Whose Knowledge?* (Ithaca: Cornell University Press, 1991); Nancy Hartsock, "The Feminist Standpoint: Developing the Ground for a Specifically Feminist Historical Materialism," in *Discovering Reality: Feminist Perspectives on Epistemology, Metaphysics, Methodology, and Philosophy of Science*, ed. Sandra Harding and Merrill Hintikka (Dordrecht: Reidel/Kluwer, 1983); Dorothy Smith, *The Everyday World as Problematic: A Sociology for Women* (Boston: Northeastern University Press, 1987), and her *The Conceptual Practices of Power: A Feminist Sociology of Knowledge* (Boston: Northeastern University Press, 1990).

5. In keeping with the emerging practice in postcolonial science and technology writings, "northern" and "southern" here replace such earlier contrasts as first and third world, west and orient, and so on. These terms are also used in the literatures discussed to refer to the conceptual frameworks arising from such locations in global political relations, though, of course, conceptual frameworks

are always cultural products, shaped by local, diverse interests, discursive resources, and so on. Such language issues were discussed in earlier chapters.

6. C. P. Snow's comments in *Two Cultures: And a Second Look* (Cambridge: Cambridge University Press, 1964) about the dangers of the continued existence of the two cultures provide a good example of the way this kind of thinking about the benefits of modern sciences for the developing societies was argued to non-scientists in the North in the late 1950s.

7. Earlier chapters reviewed postcolonial science and technology approaches that saw much of northern development policy as a continuation of imperialism by other means and that linked its failures to constitutive features of modern European sciences. See, e.g., Thomas A. Bass, *Camping with the Prince, and Other Tales of Science in Africa* (Boston: Houghton Mifflin, 1990); Lucille Brockway, *Science and Colonial Expansion: The Role of the British Royal Botanical Gardens* (New York: Academic, 1979); Andre Gunder Frank, *Capitalism and Underdevelopment in Latin America* (New York: Monthly Review Press, 1969); Susantha Goonatilake, *Aborted Discovery: Science and Creativity in the Third World* (London: Zed, 1984); Daniel R. Headrick, ed., *The Tools of Empire: Technology and European Imperialism in the Nineteenth Century* (New York: Oxford University Press, 1981); Deepak Kumar, *Science and Empire: Essays in Indian Context, 1700–1947* (Delhi, India: Anamika Prakashan and National Institute of Science, Technology, and Development, 1991); Nathan Reingold and Marc Rothenberg, *Scientific Colonialism: A Cross-Cultural Comparison* (Washington, D.C.: Smithsonian Institution Press, 1987); James E. McClellan, *Colonialism and Science: Saint Domingue in the Old Regime* (Baltimore: Johns Hopkins University Press, 1992); Charles Moraze, ed., *Science and the Factors of Inequality* (Paris: UNESCO, 1979); Ashis Nandy, ed., *Science, Hegemony, and Violence* (Delhi: Oxford, 1990); Patrick Petitjean et al., eds., *Science and Empires: Historical Studies about Scientific Development and European Expansion* (Dordrecht: Kluwer, 1992); Wolfgang Sachs, *The Development Dictionary: A Guide to Knowledge as Power* (Atlantic Highlands, N.J.: Zed, 1992); Z. Sardar, ed., *The Revenge of Athena: Science, Exploitation, and the Third World* (London: Mansell, 1988); Immanuel Wallerstein, *The Modern World System*, vol. 1 (New York: Academic, 1974). We turn shortly to the feminist elements in this discussion.

8. A few of the most influential of these northern post-positivist accounts are Nancy Cartwright, *How the Laws of Physics Lie* (New York: Oxford University Press, 1983); Anne Fausto-Sterling, *Myths of Gender: Biological Theories about Women and Men*, 2d ed. (New York: Basic, 1994); Peter Galison, *How Experiments End* (Chicago: University of Chicago Press, 1987); Peter Galison and David J. Stump, eds., *The Disunity of Science* (Stanford: Stanford University Press, 1996); Ian Hacking, *Representing and Intervening* (Cambridge: Cambridge University Press, 1983); Donna Haraway, *Primate Visions: Gender, Race, and Nature in the World of Modern Science* (New York: Routledge, 1989); Mary Hesse, *Models and Analogies in Science* (Notre Dame: University of Notre Dame Press, 1966); Evelyn Fox Keller, *Reflections on Gender and Science* (New Haven: Yale University Press, 1984), and *Secrets of Life, Secrets of Death: Essays on Language, Gender, and Science* (New York: Routledge, 1992); Bruno Latour, *Science in Action* (Cambridge: Harvard University Press, 1987), and *The Pasteurization of France* (Cambridge: Harvard University Press, 1988); Bruno Latour and Steve Woolgar, *Laboratory Life: The Social Construction of Scientific Facts* (Beverly Hills, Calif.: Sage, 1979); W. V. O. Quine, "Two Dogmas of Empiricism," in *From a Logical Point of View* (Cambridge: Harvard University Press, 1953); Sal Restivo, *Mathematics in Society and History: Sociological Inquiries* (Dordrecht: Kluwer, 1992); Richard Rorty, *Philoso-*

phy and the Mirror of Nature (Princeton: Princeton University Press, 1979); John A. Schuster and Richard R. Yeo, eds., *The Politics and Rhetoric of Scientific Method: Historical Studies* (Dordrecht: Reidel, 1986); Steven Shapin, *A Social History of Truth* (Chicago: University of Chicago Press, 1994); Steven Shapin and Simon Schaffer, *Leviathan and the Air Pump* (Princeton: Princeton University Press, 1985); Sharon Traweek, *Beamtimes and Lifetimes* (Cambridge, Mass.: MIT Press, 1988). See also Sandra Harding, *The Science Question in Feminism* (Ithaca: Cornell University Press, 1986); Sandra Harding and Merrill Hintikka, eds., *Discovering Reality: Feminist Perspectives on Epistemology, Metaphysics, Methodology, and Philosophy of Science* (Dordrecht: Reidel/Kluwer, 1983).

9. For the most influential early feminist criticism of development policies, see Ester Boserup, *Women's Role in Economic Development* (New York: St. Martin's Press, 1970). For important recent writings, see Rosi Braidotti et al., *Women, the Environment, and Sustainable Development* (Atlantic Highlands, N.J.: Zed, 1994); Irene Dankelman and Joan Davidson, *Women and Environment in the Third World: Alliance for the Future* (London: Earthscan and International Union for the Conservation of Nature, 1988); Cynthia Enloe, *Bananas, Beaches, and Bases: Making Feminist Sense of International Politics* (Berkeley: University of California Press, 1990); Wendy Harcourt, ed., *Feminist Perspectives on Sustainable Development* (London: Zed, 1994); Maria Mies, *Patriarchy and Accumulation on a World Scale: Women in the International Division of Labor* (Atlantic Highlands, N.J.: Zed, 1986); Val Plumwood, *Feminism and the Mastery of Nature* (New York: Routledge, 1993); Gita Sen and Caren Grown, *Development Crises and Alternative Visions: Third World Women's Perspectives* (New York: Monthly Review Press, 1987); Vandana Shiva, *Staying Alive: Women, Ecology, and Development* (London: Zed, 1989); Pamela Sparr, ed., *Mortgaging Women's Lives: Feminist Critiques of Structural Adjustment* (London: Zed, 1994).

10. See Mies, *Patriarchy,* and Shiva, *Staying Alive,* for clear statements of this hidden development policy.

11. Harcourt, *Feminist Perspectives.*

12. Their example is Paul Ekins and Manfred Nax-Neef, eds., *Real-Life Economics: Understanding Wealth Creation* (New York: Routledge, 1992).

13. Their example is James Robertson, *Future Wealth: A New Economics for the Twenty-first Century* (London: Cassell, 1989).

14. Their example is Sachs, *Development Dictionary.*

15. Their examples are Frederique Apffel-Marglin and Stephen Marglin, eds., *Dominating Knowledge: Development, Culture, and Resistance* (Oxford: Clarendon Press, 1990), and B. Mazlish, "The Breakdown of Connections and Modern Development," *World Development* 19:1 (1991).

16. The fullest articulation of it in the post-Kuhnian school can be found in *Disunity,* ed. Galison and Stump.

17. See Enloe, *Bananas,* and J. Ann Tickner, *Gender in International Relations: Feminist Perspectives on Achieving Global Security* (New York: Columbia University Press, 1992).

18. Though not invariably easy to resist. See, for example, Bina Agarwal's criticism, in "The Gender and Environment Debate: Lessons from India," *Feminist Studies* 18:1 (1993), of Shiva's essentialism in the latter's *Staying Alive.* The challenge is to find the level of generality appropriate to the arguments one is making.

19. The preceding chapters reviewed in further detail this more accurate and comprehensive understanding of gender relations that now prevails.

20. Such a usage threatens another form of eurocentrism by using the Euro-

pean term to refer to all such systematic knowledge, whether or not their creators conceptualized it in that way. And of course it is only a late European term; Newton and Boyle did "natural philosophy" according to their peers. The three schools of post–World War II science studies have had the effect of shifting the meanings and referents of what used to be uncontroversially and unambiguously referred to as "science" and as "ethnosciences" or "local knowledge systems." Although significant differences between modern sciences and other local knowledge systems exist, the familiar ways of marking those differences no longer are regarded as accurate or useful. These issues were discussed in an earlier chapter.

21. They are useful reasons because it is valuable to be able to see how such sciences as physics, chemistry, and theoretical biology, purportedly most immunized against carrying social values and interests, nevertheless necessarily invariably do so.

22. This is a main feature of recent social studies of science, for example, in Latour, *Pasteurization,* and Shapin and Schaffer, *Leviathan,* as cited earlier. But it is also a central focus of the "science and empires" analyses in postcolonial science studies. See, e.g., Goonatilake, *Aborted Discovery;* Nandy, *Science;* and Petitjean et al., *Science and Empires.*

23. In post-Kuhnian northern science studies, Shapin is especially interesting on this point in his *Social History.* For mathematics, see Morris Kline, *Mathematics: The Loss of Certainty* (New York: Oxford, 1980), where he points out that many of world's leading mathematicians hold that the ultimate test of the adequacy of a mathematical statement is whether it works to do what mathematicians intend it to do, which is a matter of social negotiation. See also Restivo, *Mathematics in Society,* where he proposes that mathematicians are workers, like any others, who transform the initially everyday objects into more abstract ones, and then they work on the new abstract objects to create even more abstract ones, and so forth, in processes that successively obscure the nevertheless real relations of their work to the everyday world. Social negotiations are necessary throughout such processes.

24. The analysis of this chapter appears in a somewhat different form in *Hypatia* 13:2 (1998).

8. Recovering Epistemological Resources

1. It is "internalist" because it attributes the successes of science to science's internal epistemological features—its distinctive method, standards for objectivity, and so on. In contrast, the post-internalist studies attribute the successes also in large part to larger social formations—the historical eras—within which modern scientific projects and their epistemologies co-evolved. This is not an "externalist" epistemology because science and society are co-producing each other in these newer accounts. They are inextricably intertwined as they historically *co*-evolve. See, for example, Lucile H. Brockway, *Science and Colonial Expansion: The Role of the British Royal Botanical Gardens* (New York: Academic, 1979); Bruno Latour, *The Pasteurization of France* (Cambridge: Harvard University Press, 1988); Patrick Petitjean et al., eds., *Science and Empires: Historical Studies about Scientific Development and European Expansion* (Dordrecht: Kluwer, 1992); and Steven Shapin and Simon Schaffer, *Leviathan and the Air Pump* (Princeton: Princeton University Press, 1985) for just a few examples of how this worked at significant moments in the history of northern sciences.

2. The "values and interests" framework acknowledges only the pre-1960s

liberal and Marxist concerns. It does not yet conceptualize the full extent of what historian Thomas Kuhn signaled as the "integrity of sciences with their historical era." Thomas S. Kuhn, *The Structure of Scientific Revolutions*, 2d ed. (Chicago: University of Chicago Press, 1970).

3. This chapter draws on a number of earlier writings on this topic. Its arguments for the desirability of what eventually came to be called "strong objectivity" were explored in a number of papers and edited collections in the 1970s and early 1980s and structured the argument of Sandra Harding, *The Science Question in Feminism* (Ithaca: Cornell University Press, 1986). The most extensive accounts of the notion on which this chapter directly draws are in chapter 6 of Sandra Harding, *Whose Science? Whose Knowledge?* (Ithaca: Cornell University Press, 1991); "After the Neutrality Ideal: Science, Politics, and 'Strong Objectivity,'" *Social Research* 59 (1992), 567–87, reprinted in *The Politics of Western Science*, ed. Margaret C. Jacobs (Atlantic Highlands, N.J.: Humanities, 1993); "Rethinking Standpoint Epistemology," in *Feminist Epistemologies*, ed. L. Alcoff and E. Potter (New York: Routledge, 1992); "Can Feminist Thought Make Economics More Objective?" *Feminist Economics* 1:1 (1995), 7–32; "'Strong Objectivity': A Response to the New Objectivity Question," *Synthese* 104:3 (1995), 1–19.

4. Earlier versions of this section and the next appear in "'Strong Objectivity.'"

5. N. Katherine Hayles, "Constrained Constructivism: Locating Scientific Inquiry in the Theater of Representation," in *Realism and Representation*, ed. George Levine (Madison: University of Wisconsin Press, 1993).

6. Bas Van Fraassen and Jill Sigman, "Interpretation in Science and in the Arts," in *Realism and Representation*, ed. Levine.

7. "So-called" since earlier chapters revealed how much of the scientific and technological traditions of other cultures modern European sciences had incorporated, and continue to do so today through "development" projects. Modern European science is already in this respect multicultural.

8. The phrase is a paraphrase of R. K. Kochhar, "Science in British India," in *Current Science* (India), part I: 63:11 (1992), 689–94; part II: 64:1 (1993) 55–62.

9. This issue was developed in chapters 2 and 3.

10. Or, in the nineteenth-century formulation that has left problematic residues in contemporary epistemology: "objectivity or subjectivity? . . . ," as Robert Proctor points out in *Value-Free Science? Purity and Power in Modern Knowledge* (Cambridge: Harvard University Press, 1991). Today, referring to all values and interests as subjective ones obscures the all-important difference between those that are idiosyncratically held by individuals and those that are culture-wide, such as ideologies, worldviews, and so on. Androcentrism, racism, euro-centrism, and so on, are fundamentally properties not of individuals' but of cultures' belief systems. They are examples of social, institutional, and civilizational or philosophic ethnocentrism, not of intentional or unintentional individual "prejudice." See James J. Scheurich and Michelle D. Young, "Coloring Epistemologies," *Educational Researcher* 26:3 (1997), 4–16, for this kind of analysis of social, institutional, and civilizational racism.

11. Following the practice of Richard Bernstein, *Beyond Objectivism and Relativism* (Philadelphia: University of Pennsylvania Press, 1983), Evelyn Fox Keller, *Reflections on Gender and Science* (New Haven: Yale University Press, 1984), and others, this chapter will use "objectivism" to refer to the conventional concept that takes neutrality to be a requirement for maximizing objectivity. The term "objectivity" will be reserved for the "strong objectivity," shorn of the neutrality requirement, proposed below.

12. Kuhn, *Structure*.

13. See Janice Moulton's arguments about the limitations of such a model of good thinking in "A Paradigm of Philosophy: The Adversary Method," in *Discovering Reality: Feminist Perspectives on Epistemology, Metaphysics, Methodology, and Philosophy of Science*, ed. Sandra Harding and Merrill Hintikka (Dordrecht: Reidel/Kluwer, 1983).

14. For reasons to be recounted below, claims to less falsity are preferable to those for truth or verisimilitude. See Alan Megill, "Rethinking Objectivity," *Annals of Scholarship* 8:3 (1991) for a related account of four senses of objectivity prevalent in the history of philosophy: absolute objectivity (i.e., realism), disciplinary consensus, dialectical objectivity (where objects are constructed through the interactions between subjective and objective processes), and procedural objectivity (i.e., an impersonal method). The strong objectivity developed below is different from all four, though it includes elements of both dialectical and procedural objectivity. See also Lisa Lloyd, "Science and Anti-Science: Objectivity and Its Real Enemies," in *Feminism, Science, and the Philosophy of Science*, ed. Lynn Hankinson Nelson and Jack Nelson (Boston: Kluwer, 1996), 217–59.

15. Peter Novick, *That Noble Dream: The "Objectivity Question" and the American Historical Profession* (Cambridge: Cambridge University Press, 1988), 1.

16. Ibid., 2.

17. Proctor, *Value-Free Science?* 262.

18. Ibid.

19. The standpoint epistemologies referred to here will be discussed more extensively in the next chapter.

20. Richard Levins and Richard Lewontin, "Applied Biology in the Third World," from their *The Dialectical Biologist* (Cambridge: Harvard University Press, 1988); reprinted in *The "Racial" Economy of Science*, ed. Sandra Harding (Bloomington: Indiana University Press, 1993), 315–16. Similar arguments, though considerably more skeptical about the progressive effects of early modern science on anyone but Europeans in the dominant classes, can be found in other feminist and postcolonial accounts. See, e.g., Susantha Goonatilake, *Aborted Discovery: Science and Creativity in the Third World* (London: Zed, 1984); Ashis Nandy, ed., *Science, Hegemony, and Violence: A Requiem for Modernity* (Delhi: Oxford, 1990); Ziauddin Sardar, ed., *The Revenge of Athena: Science, Exploitation, and the Third World* (London: Mansell, 1988), and the references below to the discussion of the masculinity of neutrality.

21. The account of the following few pages appeared in "After the Neutrality Ideal."

22. Even if one thinks no sciences are "pure" of all social values and interests, they can at designated historical moments be pure of particular ones. For example, particular configurations of racist and sexist assumptions entered biology during the late eighteenth and early nineteenth centuries, as historians have pointed out. See Stephen Jay Gould, *The Mismeasure of Man* (New York: Norton, 1981); Londa Schiebinger, *The Mind Has No Sex? Women in the Origins of Modern Science* (Cambridge: Harvard University Press, 1989), and *Nature's Body: Gender in the Making of Modern Science* (Boston: Beacon, 1993).

23. See Andrew Pickering, ed., *Science as Practice and Culture* (Chicago: University of Chicago Press, 1992); Joseph Rouse, *Knowledge and Power: Toward a Political Philosophy of Science* (Ithaca: Cornell University Press, 1987); Steven Shapin and Simon Shaffer, *Leviathan and the Air Pump* (Princeton: Princeton University Press, 1985). Some have said "authoritarian" science; cf. Paul Feyerabend, *Against Method* (London: New Left, 1975).

24. Robert Proctor, *Racial Hygiene: Medicine under the Nazis* (Cambridge: Harvard University Press, 1988), 290, 293.

25. Kathryn Pyne Addelson, "The Man of Professional Wisdom," in *Discovering Reality,* ed. Harding and Hintikka; Susan Bordo, *The Flight to Objectivity* (Albany: State University of NewYork Press, 1987); Ruth Hubbard, *The Politics of Women's Biology* (New Brunswick: Rutgers University, 1990); Alison Jaggar, "Love and the Emotions," in *Body/Knowledge,* ed. Susan Bordo and Alison Jaggar (New Brunswick: Rutgers University Press, 1990); Keller, *Reflections;* Genevieve Lloyd, *The Man of Reason: "Male" and "Female" in Western Philosophy* (Minneapolis: University of Minnesota Press, 1984); Catharine MacKinnon, "Feminism, Marxism, Method, and the State," *Signs* 7:3 (1982); Carolyn Merchant, *The Death of Nature: Women, Ecology, and the Scientific Revolution* (New York: Harper and Row, 1980).

26. See, e.g., Lucille H. Brockway, *Science and Colonial Expansion: The Role of the British Royal Botanical Gardens* (NewYork: Academic, 1979); Goonatilake, *Aborted Discovery;* Petitjean et al., eds., *Science and Empires; Revenge,* ed. Sardar.

27. See Donna Haraway, *PrimateVisions: Gender, Race, and Nature in theWorld of Modern Science* (NewYork: Routledge, 1989); John A. Schuster and Richard R.Yeo, eds., *The Politics and Rhetoric of Scientific Method: Historical Studies* (Dordrecht: Reidel/Kluwer, 1986); Steven Shapin, *A Social History ofTruth* (Chicago: University of Chicago Press, 1994); Shapin and Schaffer, *Leviathan.* Some analyses listed here in one or another of the post-Kuhnian, postcolonial, or feminist categories should also appear in the others; these three distinctive tendencies in post–World War II science and technology studies have often interacted and overlapped.

28. This is not to say that no other good advice for successful research is given out in "methods" courses.

29. Andrew Pickering, "Objectivity and the Mangle of Practice?" in *Annals of Scholarship,* 8:3 (1992), ed. Alan Megill.

30. Hayles, *Constrained Constructivism.*

31. National Academy of Sciences, *On Being a Scientist* (Washington, D.C.: National Academy Press, 1989), 5–6.

32. Gould, *Mismeasure,* 21–22.

33. Some might think this problem can be resolved by adding members of excluded groups into the community, or by building more conflict into scientific processes. Efforts in these directions can be helpful, but reflection on Gould's discussion makes clear that success at such strategies requires massive political changes. Won't those "included" be only the well socialized, least critical of the excluded? What kind of severe criticism should one expect to arise, and be taken seriously, from junior colleagues who are members of devalued gender, racial, ethnic, or class groups? See Patricia Hill Collins, *Black Feminist Thought* (New York: Routledge, 1991) and Moulton, "A Paradigm of Philosophy: The Adversary Method," on limitations of the adversarial model for advancing knowledge.

34. Or, sometimes, subjectivism; see note 1.

35. See, e.g., the adoption of a disabling relativism in David Bloor, *Knowledge and Social Imagery* (London: Routledge and Kegan Paul, 1977); of a subjectivist epistemology in Lorraine Code, *What Can She Know?* (Ithaca: Cornell University Press, 1991); and the strengthening of relativist epistemologies in Paul Feyerabend, "Notes on Relativism," in his *Farewell to Reason* (New York: Verso, 1987). Of course other analyses have been called relativist though they do not espouse such a position.

36. This section and the next two draw have appeared in similar forms in my "Can FeministThought Make Economics More Objective?"

37. For example, the essays in *Feminism and Methodology*, ed. Sandra Harding (Bloomington: Indiana University Press, 1987); Hubbard, *Politics*; Keller, *Reflections*.

38. Bordo, *Flight*, 451.

39. Lloyd, *Man of Reason*.

40. Genevieve Lloyd, "Maleness, Metaphor, and the 'Crisis' of Reason," in *A Mind of One's Own: Feminist Essays on Reason and Objectivity*, ed. Louise Antony and Charlotte Witt (Boulder: Westview, 1993).

41. Keller, *Reflections*, 75. She cites Karen Horney's quotation from Simmel in Horney's, "The Flight from Womanhood," in *Women and Analysis*, ed. J. Strouse (New York: Dell 1975), 199–215.

42. Catharine MacKinnon, "Feminism, Marxism, Method, and the State: Toward Feminist Jurisprudence," *Signs* 8:4 (1983), 658.

43. See Phyllis Rooney, "Recent Work in Feminist Discussions of Reason," *American Philosophical Quarterly* 31:1 (1994), 1–21, for a thorough review of the U.S. feminist literature on the related issue of rationality.

44. Harry Brod, ed., *The Making of Masculinities*, 2d ed. (New York: Allen and Unwin, 1995).

45. We should at least consider the possibility that while there may well be lots of other good reasons to balk at such a delinking, at least one source of resistance to it may be fear of the loss of the masculinity of science, philosophy of science, and epistemology. Philosophies of objectivity, rationality, and science that do not proclaim their distance from, their transcendence of, their unaccountability to, the messy moral and political demands of the social order may not be as attractive to people seeking affirmations of their masculinity through practicing in such fields. Objectivity without neutrality may for them have the odor of social work.

46. At least as imagined by the playwright Bertolt Brecht in *The Life of Galileo*, trans. Howard Brenton (London: Eyre Methuen, 1981).

47. Sandra Harding, ed., *Can Theories Be Refuted? Essays on the Duhem-Quine Thesis* (Dordrecht: Reidel/Kluwer, 1976); W. V. O. Quine, "Two Dogmas of Empiricism," in *From a Logical Point of View* (Cambridge: Harvard University Press, 1953).

48. Cf. Peter Galison, *How Experiments End* (Chicago: University of Chicago Press, 1987); Kuhn, *Structure*; Feyerabend, *Against Method*. Here is where scientific communities especially benefit from looking beyond their currently legitimated borders for guidance.

49. This is pointed out by Hayles in "Constrained Constructivism" and by Van Fraassen and Sigman in "Interpretation."

50. Gould, *Mismeasure*.

51. See Sandra Harding, "Are Truth Claims Dysfunctional?" in *Philosophy of Language*, ed. Andrea Nye (New York: Blackwell, 1998).

9. Borderlands Epistemologies

1. Langdon Winner made this point in his "The Gloves Come Off: Shattered Alliances in Science and Technology Studies," *Social Text* 46–47 (1996), 81–92.

2. Thomas S. Kuhn, *The Structure of Scientific Revolutions*, 2d ed. (Chicago: University of Chicago Press, 1970), 1. Kuhn was by no means the only stimulant to the new science and technology studies, as has been discussed in earlier chapters. Historical changes in global social and economic relations; the rise of various science and technology social movements focussed on, for example,

antimilitarism, environmentalism, labor conditions, black health and women's health; the rise in racist and sexist deterministic biological theories; the increasing accomplishments of the social history that had emerged in the 1930s; and new projects already under way in the sociology, ethnography, and philosophy of science were some of the other changes that helped to fuel the post–World War II science and technology studies. Kuhn's book is used here simply as a convenient marker for the take-off of these new approaches.

3. See, e.g., W. V. O. Quine, "Two Dogmas of Empiricism," in his *From a Logical Point of View* (Cambridge: Harvard University Press, 1953). See chapter 15 of Peter Novick, *That Noble Dream: The "Objectivity Question" and the American Historical Profession* (Cambridge: Cambridge University Press, 1988) for a good overview of the historical changes and intellectual ferment of the post–World War II period that produced and surrounded the emerging "epistemological crisis of the modern West," including major streams of the science and technology studies examined here. Richard Bernstein's *Beyond Objectivism and Relativism* (Philadelphia: University of Pennsylvania Press, 1983) and Richard Rorty's *Philosophy and the Mirror of Nature* (Princeton: Princeton University Press, 1979) were two immensely influential books by philosophers that diagnosed how "the crisis" was emerging across broad swaths of European and North American natural and social sciences and their philosophies.

4. As noted in earlier chapters, these three streams of science and technology studies are by no means entirely separate; they have interacted and their arguments often coincide. Some writers and writings, such as the latest stages of the gender and sustainable development debates and Donna Haraway's work, for example, clearly have been shaped by all three tendencies. Nevertheless, most of the post–World War II science and technology writings have not drawn on this full range of analyses; they are still constrained by pre-Kuhnian, eurocentric and/or androcentric understandings of modern science.

5. The idea is Michel Foucault's. See, for example, his *Power/Knowledge: Selected Interviews and Other Writings, 1972–77*, trans. Colin Gordon, Leo Marshall, John Mepham, and Kate Soper (New York: Random House, 1980).

6. As I have argued in a somewhat different context, in this respect they are as conservative as the internalists, who agree that there is no reasonable epistemological alternative to their own program. See my chapter 7, "Feminist Epistemology in after the Enlightenment," in *Whose Science? Whose Knowledge?* (Ithaca: Cornell University Press, 1991).

7. Interestingly, historians of science evidently have felt less need to enter this fray, though their illuminating accounts of the history of science as social and cultural history, as well as intellectual history, have in fact usefully created a lot of the trouble. Kuhn's *Structure*, for example, explicitly framed an epistemological account as challenging the dominant epistemology of modern science. Peter Galison and David J. Stump's *The Disunity of Science* (Stanford: Stanford University Press, 1996) presents many historians and philosophers proposing not just to fix up or reject internalist epistemologies, but rather to radically transform them.

8. My use of the term *borderlands* in this chapter is indebted to Gloria Anzaldua's *Borderlands/La Frontera* (San Francisco: Spinsters/Aunt Lute Press, 1987). I have discussed standpoint epistemologies in many earlier publications. For some of the most extensive of these discussions, see *The Science Question in Feminism;* (Ithaca: Cornell University Press, 1986); *Whose Science? Whose Knowledge?; "Rethinking Standpoint Epistemology," in *Feminist Epistemologies*, ed. L. Alcoff and E. Potter (New York: Routledge, 1992); and "How Social Disad-

vantage Creates Epistemic Advantage," in *Social Theory and Sociology*, ed. Stephen Turner (New York: Blackwell, 1996).

9. The resources of such culturally specific standpoints were explored especially in chapter 4. Early feminist standpoint theories were explicitly concerned with the effects on knowledge of the *politics* of men's and women's *culturally distinctive* activities. How have the systematic patterns of knowledge and ignorance in the modern West reflected the exclusion of women from the conceptualization of sociological or philosophical problems, for example, and the relative absence of men from the childcare, household labor, and emotional labor which had been assigned primarily to women? See, e.g., Nancy Hartsock, "The Feminist Standpoint: Developing the Ground for a Specifically Feminist Historical Materialism," in *Discovering Reality: Feminist Perspectives on Epistemology, Metaphysics, Methodology, and Philosophy of Science,* ed. Sandra Harding and Merrill Hintikka (Dordrecht: Reidel/Kluwer, 1983); Alison Jaggar, *Feminist Politics and Human Nature* (Totowa, N.J.: Rowman and Allenheld, 1983), chapter 11; Hilary Rose, "Hand, Brain, and Heart: A Feminist Epistemology for the Natural Sciences," *Signs* 9:1 (1983); Dorothy Smith, *The Everyday World as Problematic: A Sociology for Women* (Boston: Northeastern University Press, 1987), and *The Conceptual Practices of Power: A Feminist Sociology of Knowledge* (Boston: Northeastern University Press, 1990). Smith's essays collected in these volumes had been appearing since the mid 1970s. See also my discussions of standpoint theories in *The Science Question in Feminism* (Ithaca, Cornell University Press, 1986), and in *Whose Science? Whose Knowledge?*

10. Of course, there were many problems with the way the proletarian standpoint was conceptualized. After Lukacs's work on it, feminist theorists were the next to try to use the resource of a specifically Marxian understanding of the relationship between "doing" and knowing to develop an epistemology. Of course Foucault had meanwhile been exploring power/knowledge relations in ways that lead readers to the disavowal of any epistemological projects at all. See his *Power/Knowledge*. Fredric Jameson has argued that the feminist standpoint theorists are the only contemporary thinkers currently working explicitly with the Marxian epistemological legacy. See George Lukacs, *History and Class Consciousness* (Cambridge, Mass.: MIT Press, 1971); Jameson, "*History and Class Consciousness* as an 'Unfinished Project,'" *Rethinking Marxism* 1 (1988), 49–72; F. Engels, "Socialism: Utopian and Scientific," in *The Marx and Engels Reader,* ed. R. Tucker (New York: Norton, 1972).

11. "Post-Marxisms" come after Marxism and draw on its resources, among other sources of their projects. From this perspective, major strains of feminism, antiracism, postcolonialism, contemporary class-based analyses, the environmental movement, antimilitarism, and so on, are all "post-Marxisms." This is a different relationship to Marxism than "anti-Marxists" have, and anti-Marxist tendencies can also be found in all of these movements. Moreover, U.S. intellectual life is much less post-Marxist than is such life in Europe, since Marxism has never been treated as part of an important intellectual legacy in the United States, as it has in Europe. U.S. intellectual life has never had available enough Marxian thought to get to post-Marxism. This situation prepares readers differently on the two continents for understanding contemporary standpoint writings and gives these writings different meanings on the two sides of the Atlantic.

12. Cf., for example, Edward Said, *Orientalism* (New York: Pantheon, 1978); Samir Amin, *Eurocentrism* (New York: Monthly Review Press, 1989); Monique Wittig, "The Straight Mind," *Feminist Issues* 1:1 (1980); and many of the postcolonial writings cited in this book.

13. Patricia Hill Collins's account is particularly interesting in this regard since her use of slave narratives, oral histories, interviews, diaries, autobiographies, the lyrics of blues, and other records of African American women's voices enables the reader to hear standpoint arguments again and again in the thinking of these political activists and everyday women. See her *Black Feminist Thought: Knowledge, Consciousness, and the Politics of Empowerment* (New York: Routledge, 1991). Another of the many places where spontaneous standpoint arguments consistently appear is in the critiques of development literature, especially in the case studies. See, e.g., Wendy Harcourt, ed., *Feminist Perspectives on Sustainable Development* (London: Zed, 1994).

14. The heading is borrowed from Smith's book, *Conceptual Practices.*

15. Hartsock, "The Feminist Standpoint."

16. Ibid., 159. Hartsock's use of the term "real relations" has suggested to some readers that she and other standpoint theorists are hopelessly mired in an epistemology and metaphysics that have been discredited by social constructionists. This judgment fails to appreciate the way standpoint theories reject *both* pure realist and pure social constructionist epistemologies and metaphysics. Donna Haraway is particularly good on this issue: "Situated Knowledges: The Science Question in Feminism and the Privilege of Partial Perspectives," in *Simians, Cyborgs, and Women* (New York: Routledge, 1991).

17. See, e.g., Harry Brod, ed., *The Making of Masculinity: The New Men's Studies,* 2d ed. (New York: Allen and Unwin, 1987); Sandra Harding, "Can Men Be Subjects of Feminist Thought?" in *Men Doing Feminism,* ed. Tom Digby III (New York: Routledge, 1997).

18. Of course body work and emotional work also require head work—contrary to the long history of sexist, racist, and class biased views. See, e.g., Sara Ruddick, *Maternal Thinking* (Boston: Beacon, 1989). And the kind of head work required in administrative and managerial work—what Smith means by "ruling"—also involves distinctive body and emotional work, though it is not acknowledged as such. Think of how much of early childhood education of middle-class children is really about internalizing a certain kind of (gender specific) regulation of bodies and emotions.

19. See the very similar accounts, independently produced, in Hartsock's "Feminist Standpoint" and Rose's "Hand, Brain, and Heart." Feminist standpoint theory was an idea whose time had arrived in Canada, the United States, and the United Kingdom.

20. See, e.g., Amin, *Eurocentrism;* J. M. Blaut, *The Colonizer's Model of the World* (New York: Guilford Press, 1993); Vandana Shiva, *Staying Alive: Women, Ecology, and Development* (London: Zed, 1989).

21. I specify "prefeminist empiricism" here since some of the feminist philosophical empiricists, such as Helen Longino, who have always understood the importance of progressive politics for "eliminating bias" from purportedly value-free claims, have recently begun to permit somewhat more expanded contributions for such politics, while nevertheless drawing a firm line between their projects and the standpoint centering of relations between politics and knowledge that directs us all to "start off thought from marginal lives" in order to gain more accurate and comprehensive accounts. See her *Science as Social Knowledge* (Princeton: Princeton University Press, 1990).

22. Harding, *Whose Science? Whose Knowledge?*, 121–33.

23. Marcia Millman and Rosabeth Ross Kanter, "Introduction to *Another Voice: Feminist Perspectives on Social Life and Social Science,"* in *Feminism and Methodology,* ed. Sandra Harding (Bloomington: Indiana University Press, 1987), 32.

24. Patricia Hill Collins has discussed this in *Black Feminist.*

25. "Outsider within" is Patricia Hill Collins's phrase; see her *Black Feminist Thought.* As indicated above, *Borderlands* is Gloria Anzaldua's term.

26. See my "Why Has the Sex/Gender System Emerged into Visibility Only Now?" in *Discovering Reality: Feminist Perspectives on Epistemology, Metaphysics, Methodology, and Philosophy of Science,* ed. Sandra Harding and Merrill Hintikka (Dordrecht: Reidel, 1983).

27. Blaut, *Colonizer's Model.*

28. See, for example, Donna Haraway, *Primate Visions: Gender, Race, and Nature in the World of Modern Science* (New York: Routledge, 1989); Edward Said, *Orientalism.*

29. Feminist standpoint ambiguities and ambivalences about the role to be assigned to women's experiences have been the topic of innumerable discussions. See, for example, Rosemary Hennessy, *Feminist Materialism and the Politics of Discourse* (New York: Routledge, 1993); and many discussions in sociology journals of Smith's work in particular. However, one strain throughout standpoint theory's history within feminism, more strongly emphasized in some writings than in others, has been that women's experiences are themselves generated from within discourses—prevailing, or subjugated, or newly constructed through feminisms. Neither women's experiences nor their subjectivities are constituted prior to "the social." Accessible discussions of this topic more generally can be found in Chris Weedon, *Feminist Practice and Poststructuralist Theory* (Cambridge, Mass.: Blackwell, 1987).

30. Some of these accounts will be more complex than others. For example, while class origins do stick to one, class identity is more crossable than race or gender—at least in the contemporary United States.

31. Dorothy Smith made this claim in her "Comment on Harding," *American Philosophical Association Newsletter on Feminism and Philosophy* 88:3 (1989).

32. Quine, "Two Dogmas."

10. Dysfunctional Universality Claims?

1. As noted in earlier chapters, there are also postcolonial reasons to resist subsuming all cultures' traditions of systematic knowledge about themselves and the world around them under what the West has in the last century or so referred to as "science." (Even in the West, the term is a recent one, since Galileo, Newton, and Boyle's work was referred to as "natural philosophy.") After all, why should other cultures' projects have to be named in European terms in order to be taken seriously by Europeans?

2. Compare, e.g., Susan Bordo, *The Flight to Objectivity: Essays on Cartesianism and Culture* (Albany: State University of New York Press, 1987); Alison Jaggar, "Love and Knowledge: Emotions in Feminist Epistemology," in *Gender/Body/ Knowledge,* ed. Susan Bordo and Alison Jaggar (New Brunswick: Rutgers University Press, 1989); Evelyn Fox Keller, *Reflections on Gender and Science* (New Haven: Yale University Press, 1984); Genevieve Lloyd, *The Man of Reason: "Male" and "Female" in Western Philosophy* (Minneapolis: University of Minnesota Press, 1984); Phyllis Rooney, "Recent Work in Feminist Discussions of Reason," *American Philosophical Quarterly* 31:1, 1–21.

3. Compare, e.g., Susantha Goonatilake, "The Voyages of Discovery and the Loss and Rediscovery of the 'Other's' Knowledge," *Impact of Science on Society* no. 167 (1992), 241–64; R. K. Kochhar, "Science in British India," parts I and II, *Current Science* 63:11 (1992–93), 689–94; and 64:1 (1992–93) 55–62 (India); Joseph Needham, *The Grand Titration: Science and Society in East and West* (Toronto:

University of Toronto Press, 1969); Patrick Petitjean et al., eds., *Science and Empires: Historical Studies about Scientific Development and European Expansion* (Dordrecht: Kluwer, 1992); Ziauddin Sardar, ed., *The Revenge of Athena: Science, Exploitation, and the Third World* (London: Mansell, 1988).

4. For illuminating recent accounts of the history and current scientific support for these unity of science claims, see John Dupre, *The Disorder of Things: Metaphysical Foundations for the Disunity of Science* (Cambridge: Harvard University Press, 1993); and *The Disunity of Science*, ed. Peter Galison and David Stump (Stanford: Stanford University Press, 1996). For one exploration of problems with the unity of science claims from a postcolonial perspective, see David J. Hess, *Science and Technology in a Multicultural World* (New York: Columbia University Press, 1995). For an older harbinger of these arguments, see Patrick Suppes, "The Plurality of Science," *Philosophy of Science Association 1978*, vol. 2, ed. P. Asquith and I. Hacking (East Lansing: Philosophy of Science Association, 1978). Chapter 4 above, "Cultures as Toolboxes for Sciences and Technologies," examined what it is about nature and social relations that insures that science and technology inevitably and desirably must be plural. Note that the universality claim in its unity of science form, as well as in other forms, asserts the uniquely maximal reliability of scientific claims (often expressed in terms of their truth, or in terms of the "fact that science works") and the unique validity of sciences' logic of research and explanation that produced them. The focus here will be primarily on the validity claim since it is sciences' logic of research and explanation that is thought responsible for its production of empirically reliable claims.

5. Such insights are the beginning of the development of standpoint epistemologies—only the beginning, not the end, since these insights express "identity epistemologies" while standpoint epistemologies center not socially unmediated experience but distinctive kinds of critically and dialogically achieved discourses as generators of knowledge. See chapters 8 and 9, and, e.g., Patricia Hill Collins, *Black Feminist Thought: Knowledge, Consciousness, and the Politics of Empowerment* (New York: Routledge, 1991).

6. Thanks to Val Plumwood for pointing out to me this fourth assumption in the unity of science thesis.

7. Mario Biagioli, *Galileo Courtier* (Cambridge: Harvard University Press, 1993); Bruno Latour, *The Pasteurization of France* (Cambridge: Harvard University Press, 1988); Robert Proctor, *Cancer Wars: How Politics Shapes What We Know and Don't Know about Cancer* (Boston: Basic, 1995).

8. Compare, e.g., John A. Schuster and Richard R. Yeo, eds., *The Politics and Rhetoric of Scientific Method: Historical Studies* (Dordrecht: Reidel, 1986); Steven Shapin and Simon Schaffer, *Leviathan and the Air Pump* (Princeton: Princeton University Press, 1985).

9. This was a major point of Thomas Kuhn's *The Structure of Scientific Revolutions*, 2d ed. (Chicago: University of Chicago Press, 1970) and the outpouring of subsequent histories, sociologies, ethnographies, and philosophies of science and technology that followed it. For the demise of the idea of "crucial experiments," see also Sandra Harding, ed., *Can Theories Be Refuted? Essays on the Duhem-Quine Thesis* (Dordrecht: Kluwer/Reidel, 1976).

10. Galison, "Introduction," in *Disunity*, ed. Galison and Stump, 14–15.

11. Compare, e.g., Helen Watson-Verran and David Turnbull, "Science and Other Indigenous Knowledge Systems," in *Handbook of Science and Technology Studies*, ed. S. Jasanoff, G. Markle, T. Pinch, and J. Petersen (Thousand Oaks, Calif.: Sage, 1995), 115–39.

12. Compare David Bloor, *Knowledge and Social Imagery* (London: Routledge

and Kegan Paul, 1977); George Gheverghese Joseph, *The Crest of the Peacock: Non-European Roots of Mathematics* (New York: I. B. Tauris, 1991); Sal Restivo, *Mathematics in Society and History: Sociological Inquiries* (Dordrecht: Kluwer, 1992). Arthur B. Powell and Marilyn Frankenstein, eds., *Ethnomathematics: Challenging Eurocentrism in Mathematics Education* (Albany: State University of New York Press, 1997).

13. Compare Ted J. Kaptchuk, *The Web That Has No Weaver: Understanding Chinese Medicine* (New York: Congdon and Weed, 1983).

14. Galison, *Disunity*, 5.

15. Ibid., 3–8.

16. Weinberg's original argument is in *Dreams of a Final Theory* (New York: Pantheon, 1993). A later, more measured, statement can be found in his discussion of Maxwell's equations in "Sokal's Hoax," *New York Review of Books* (8 Aug. 1996), 13.

17. Ian Hacking, "The Disunities of the Sciences," in *Disunity*, ed. Galison and Stump, 52.

18. A. C. Crombie, *Styles of Scientific Thinking in the European Tradition* (London: Duckworth, 1994); Hacking, in *Disunity*, ed. Galison and Stump.

19. Hacking, in *Disunity*, ed. Galison and Stump, 68.

20. Bloor, *Knowledge*; Morris Kline, *Mathematics: The Loss of Certainty* (New York: Oxford, 1980); Restivo, *Mathematics*.

21. These issues were discussed in earlier chapters. In addition to the citations against the unique universality of modern science provided above, see, e.g., N. Katherine Hayles, "Constrained Constructivism: Locating Scientific Inquiry in the Theater of Representation," in *Realism and Representation*, ed. George Levine (Madison: University of Wisconsin Press, 1993); Bas Van Fraassen and Jill Sigman, "Interpretation in Science and in the Arts," in *Realism*, ed. Levine.

22. See the notes to earlier chapters, e.g., J. M. Blaut, *1492: The Debate on Colonialism, Eurocentrism, and History* (Trenton, N.J.: Africa World Press, 1992); Lucille Brockway, *Science and Colonial Expansion: The Role of the British Royal Botanical Gardens* (New York: Academic, 1979); Goonatilake, "Voyage"; James E. McClellan *Colonialism and Science: Saint Domingue in the Old Regime* (Baltimore: Johns Hopkins University Press, 1992); Petitjean, *Empires*; Nathan Reingold and Marc Rothenberg, eds., *Scientific Colonialism* (Washington, D.C.: Smithsonian Institution Press, 1987); *Revenge*, ed. Sardar.

23. See Wolfgang Sachs, ed., *The Development Dictionary: A Guide to Knowledge as Power* (Atlantic Highlands, N.J.: Zed, 1992; and many of the essays in *Revenge*, ed. Sardar. See Third World Network, "Modern Science in Crisis: A Third World Response," in *Revenge*, ed. Sardar, as a separate monograph published by the Third World Network (Penang, Malaysia, 1988), and reprinted in Sandra Harding, ed., *The "Racial" Economy of Science: Toward a Democratic Future* (Bloomington: Indiana University Press, 1993).

24. See Blaut, *Debate*; Boris Hessen, *The Economic Roots of Newton's Principia* (New York: Howard Fertig, 1970).

25. Joseph, *Crest*.

26. J. M. Blaut, *The Colonizer's Model of the World: Geographical Diffusionism and Eurocentric History* (New York: Guilford Press, 1993).

27. Michael Adas, *Machines as the Measure of Man* (Ithaca: Cornell University Press, 1989).

28. Susantha Goonatilake, *Aborted Discovery: Science and Creativity in the Third World* (London: Zed, 1984); Ashis Nandy, ed., *Science, Hegemony, and Violence: A Requiem for Modernity* (Delhi: Oxford University Press, 1990); Sachs, *Develop-*

ment; Vandana Shiva, *Staying Alive: Women, Ecology, and Development* (London: Zed, 1989); *Revenge,* ed. Sardar.

29. Compare Tom Patterson's arguments that the concept "nature" has a class history. It was persistently introduced by protocapitalist "outside experts" (often groups in their own society) in their struggles with peasants and/or farmers over who would have the power to decide how land was to be used. Tom Patterson, "Nature: The Shadow of Civilization," (forthcoming), and his *Inventing Western Civilization* (New York: Monthly Review Press, 1997). See also the issues about "nature" raised in chapters 5 and 6.

30. See chapter 7 for related discussion of the dependency of modern, masculinized forms of knowledge on purportedly premodern women's forms. Maria Mies makes similar arguments in *Patriarchy and Accumulation on a World Scale: Women in the International Division of Labor* (Atlantic Highlands, N.J.: Zed, 1986).

31. Bruno Latour, *Science in Action* (Cambridge: Harvard University Press, 1987).

32. Watson-Verran and Turnbull, "Science," 117.

33. Gilles Deleuze and Felix Guattari, *A Thousand Plateaus: Capitalism and Schizophrenia* (Minneapolis: University of Minnesota Press, 1987).

34. David Turnbull, "Local Knowledge and Comparative Scientific Traditions," *Knowledge and Policy* 6:3/4 (1993), 29.

35. See, e.g., Carolyn Wood Sherif, "Bias in Psychology," in *Feminism and Methodology,* ed. Sandra Harding (Bloomington: Indiana University Press, 1987); Ruth Hubbard, *The Politics of Women's Biology* (New Brunswick: Rutgers University Press, 1990); Anne Fausto-Sterling, *Myths of Gender: Biological Theories about Women and Men,* 2d ed. (New York: Basic, 1994).

36. See, e.g., Thomas A. Bass, *Camping with the Prince, and Other Tales of Science in Africa* (Boston: Houghton Mifflin, 1990), and Vandana Shiva, *Staying Alive: Women, Ecology, and Development* (London: Zed, 1989).

37. John Dupre, "Metaphysical Disorder and Scientific Disunity," in *Disunity,* ed. Galison and Stump, 115. See also Dupre, *Disorder.*

38. Dupre, "Metaphysical," 115.

39. Dupre, "Metaphysical," also makes this point about resistance to feminist criticisms (116). Cf. Sandra Harding, *The Science Question in Feminism* (Ithaca: Cornell University Press, 1986).

40. Similar criticisms have been made of the idea that the sciences do and should make truth claims. For one review of the issues, see "Are Truth Claims Dysfunctional?" by Sandra Harding, in *Philosophy of Language: The Big Questions,* ed. Andrea Nye (New York: Blackwell, 1998).

11. Robust Reflexivity

1. I suggested a stance of "robust reflexivity" in *Whose Science? Whose Knowledge?* (Ithaca: Cornell University Press, 1991), 149–50, 161–63.

2. See, e.g., Joseph Rouse, "Feminism and the Social Construction of Scientific Knowledge," in *Feminism, Science, and the Philosophy of Science,* ed. Lynn Hankinson Nelson and Jack Nelson (Dordrecht: Kluwer, 1996).

BIBLIOGRAPHY

Abel, Emily K., and Carole H. Browner. 1998. "Selective Compliance with Biomedical Authority and the Uses of Experiential Knowledge." In *Pragmatic Women and Body Politics,* ed. M. Lock and P. Kaufert. Cambridge: Cambridge University Press.

Acker, J. 1992. "Gendered Institutions: From Sex Roles to Gendered Institutions." *Contemporary Sociology* 21, 565–69.

Adas, Michael. 1989. *Machines As the Measure of Man.* Ithaca: Cornell University Press.

Addelson, Kathryn Pyne. 1983. "The Man of Professional Wisdom." In *Discovering Reality,* ed. Sandra Harding and Merrill Hintikka. Dordrecht: Reidel.

Agarwal, Bina. 1993. "The Gender and Environment Debate: Lessons from India." *Feminist Studies* 18:1.

Alexander, M. Jacqui, and Chandra Talpade Mohanty, eds. 1997. *Feminist Genealogies, Colonial Legacies, Democratic Futures.* New York: Routledge.

Alvares, Claude. 1990. "Science, Colonialism, and Violence: A Luddite View." In *Science, Hegemony and Violence,* ed. A. Nandy. Delhi: Oxford.

Amin, Samir. 1989. *Eurocentrism.* New York: Monthly Review Press.

Antony, Louise, and Charlotte Witt. 1993. *A Mind of Her Own: Feminist Essays on Reason and Objectivity.* Boulder: Westview Press.

Anzaldua, Gloria. 1981. *Borderlands/La Frontera.* San Francisco: Spinsters/Aunt Lute.

———, ed. 1990. *Making Face/Making Soul.* San Francisco: Aunt Lute Foundation Press.

———, and Cherrie Moraga, eds. 1981. *This Bridge Called My Back: Writings by Radical Women of Color.* New York: Kitchen Table Press.

Arditti, R., et al., eds. 1980. *Science and Liberation.* Boston: South End Press.

Ashcroft, Bill, Gareth Griffiths, and Helen Tiffin. 1995. *The Post-Colonial Studies Reader.* New York: Routledge.

Bajaj, Jatinder K. 1988. "Francis Bacon, the First Philosopher of Modern Science: A Non-Western View." In *Science, Hegemony, and Violence,* ed. A. Nandy. Delhi: Oxford.

Bandyopadhyay, J., and V. Shiva. 1988. "Science and Control: Natural Resources and Their Exploitation." In *The Revenge of Athena,* ed. Z. Sardar. London: Mansell.

Barinaga, Marcia. 1993. "Is There a 'Female Style' in Science?" *Science* 260, April, 384–91.

Bass, Thomas A. 1990. *Camping with the Prince, and Other Tales of Science in Africa.* Boston: Houghton Mifflin.

Belenky, Mary F., B. M. Clinchy, N. R. Goldberger, and J. M. Tarule. 1997. *Women's Ways of Knowing: The Development of Self, Voice, and Mind.* 2d ed. New York: Basic.

Berman, Morris. 1981. *The Reenchantment of the World.* Ithaca: Cornell University Press.

Bernal, Martin. 1987. *Black Athena: The Afroasiatic Roots of Classical Civilization.* Vol. I. New Brunswick: Rutgers University Press.

Bernstein, Richard. 1983. *Beyond Objectivism and Relativism.* Philadelphia: University of Pennsylvania Press.

Biagioli, Mario. 1993. *Galileo Courtier: The Practice of Science in the Culture of Absolutism.* Chicago: University of Chicago Press.

Birke, Lynda. 1986. *Women, Feminism, and Biology.* New York: Methuen.

Blaut, J. M. 1993. *The Colonizer's Model of the World: Geographical Diffusionism and Eurocentric History.* New York: Guilford Press.

Bleier, Ruth. 1984. *Science and Gender: A Critique of Biology and Its Theories on Women.* New York: Pergamon Press.

Bloor, David. 1977. *Knowledge and Social Imagery.* London: Routledge and Kegan Paul.

Bordo, Susan. 1987. *The Flight to Objectivity: Essays on Cartesianism and Culture.* Albany: State University of New York Press.

Boserup, Esther. 1970. *Women's Role in Economic Development.* New York: St. Martin's Press.

Boston Women's Health Collective. 1970. *Our Bodies, Ourselves.* Boston: New England Free Press. (Later editions: Random House).

Braidotti, Rosi, et al. 1994. *Women, the Environment, and Sustainable Development.* Atlantic Highlands, N.J.: Zed.

Bray, Francesco. 1996. "Eloge for Joseph Needham." *Isis* 87:2. 312–17.

Brighton Women and Science Group. 1980. *Alice through the Microscope: The Power of Science over Women's Lives.* London: Virago.

Brockway, Lucille H. 1979. *Science and Colonial Expansion: The Role of the British Royal Botanical Gardens.* New York: Academic Press.

Brod, Harry, ed. 1995. *The Making of Masculinities: The New Men's Studies.* 2d ed. New York: Allen and Unwin.

Butler, Judith. 1990. *Gender Trouble: Feminism and the Subversion of Identity.* New York: Routledge.

Carson, Rachel. 1962. *Silent Spring.* Harmondsworth, U.K.: Penguin.

Cartwright, Nancy. 1983. *How the Laws of Physics Lie.* New York: Oxford University Press.

————. 1989. *Nature's Capacities and Their Measurement.* Oxford: Clarendon Press.

Cetina, Karin Knorr. 1981. *The Manufacture of Knowledge: An Essay on the Constructivist and Contextual Nature of Knowledge.* New York: Pergamon.

Chodorow, Nancy. 1978. *The Reproduction of Mothering.* Berkeley: University of California Press.

Cockburn, Cynthia. 1985. *Machinery of Dominance: Women, Men, and Technical Know-How.* London: Pluto Press.

Code, Lorraine. 1991. *What Can She Know?* Ithaca: Cornell University Press.

Collins, Patricia Hill. 1991. *Black Feminist Thought: Knowledge, Consciousness, and the Politics of Empowerment.* New York: Routledge.

Crosby, Alfred. 1972. *The Columbian Exchange: Biological and Cultural Consequences of 1492.* Westport: Greenwood.

————. 1987. *Ecological Imperialism: The Biological Expansion of Europe.* Cambridge: Cambridge University Press.

Crozet, Pascal. 1994. "Scientific Language and National Identity: Egypt from the Nineteenth Century." Paper presented at the at the Twentieth Century Science: Beyond the Metropolis Conference, co-sponsored by UNESCO and ORSTOM, Paris, 20 September 1994.

Dankelman, Irene, and Joan Davidson. 1988. *Women and Environment in the Third World: Alliance for the Future.* London: Earthscan Publications and the International Union for the Conservation of Nature.

Davis, Angela. 1983. *Women, Race, and Class.* New York: Random House.

———. 1990. *Women, Culture, Politics.* New York: Random House.

Diop, Cheikh Anta. 1974. *The African Origin of Civilization: Myth or Reality?* Trans. M. Cook. Westport: L. Hill.

Dupre, John. 1993. *The Disorder of Things: Metaphysical Foundations for the Disunity of Science.* Cambridge: Harvard University Press.

———. 1996. "Metaphysical Disorder and Scientific Disunity." In *The Disunity of Science,* ed. Peter Galison and David J. Stump. Stanford: Stanford University Press.

Easlea, Brian. 1981. *Science and Sexual Oppression.* London: Weidenfeld and Nicolson.

———. 1980. *Witch-hunting, Magic, and the New Philosophy.* Brighton, U.K.: Harvester Press.

———. 1983. *Fathering the Unthinkable: Masculinity, Scientists, and the Nuclear Arms Race.* London: Pluto Press.

Ehrenreich, Barbara, and Deirdre English. 1973. *Witches, Midwives and Healers.* Old Westbury: Feminist Press.

Engels, Friedrich. 1972. "Socialism: Utopian and Scientific." In *The Marx and Engels Reader,* ed. R. Tucker. New York: Norton.

Enloe, Cynthia. 1990. *Bananas, Beaches, and Bases: Making Feminist Sense of International Politics.* Berkeley: University of California Press.

Fausto-Sterling, Anne. (1985) 1994. *Myths of Gender: Biological Theories about Women and Men.* New York: Basic.

Ferber, Marianne A., and Julie A. Nelson, eds. 1993. *Beyond Economic Man: Feminist Theory and Economics.* Chicago: University of Chicago Press.

Feyerabend, Paul. 1975. *Against Method.* London: New Left.

Forman, Paul. 1987. "Behind Quantum Electronics: National Security as Bases for Physical Research in the U.S., 1940–1960." *Historical Studies in Physical and Biological Sciences* 18.

Foucault, Michel. 1980. *Power/Knowledge: Selected Interviews and Other Writings, 1972–77.* Trans. Colin Gordon, Leo Marshall, John Mepham, and Kate Soper. New York: Random House.

Frake, C. 1962. "The Ethnographic Study of Cognitive Systems." In *Anthropology and Human Behaviour,* ed. T. Gladwin. Washington, D.C.: Anthropology Society of Washington.

Frank, Andre Gunder. 1969. *Capitalism and Underdevelopment in Latin America.* New York: Monthly Review Press.

Fuller, Steve. 1992. "Social Epistemology and the Research Agenda of Science Studies." In *Science as Practice and Culture,* ed. Andrew Pickering. Chicago: University of Chicago Press.

Galison, Peter. 1987. *How Experiments End.* Chicago: University of Chicago Press.
————, and David J. Stump, eds. 1996. *The Disunity of Science.* Stanford: Stanford University Press.
Gilligan, Carol. 1982. *In a Different Voice.* Cambridge: Harvard University Press.
Goldberger, Nancy, et al. 1996. *Knowledge, Difference, and Power: Essays Inspired by Women's Ways of Knowing.* New York: Basic.
Goonatilake, Susantha. 1984. *Aborted Discovery: Science and Creativity in the Third World.* London: Zed.
————. 1988. "A Project for Our Times." In *The Revenge of Athena,* ed. Z. Sardar. London: Mansell.
————. 1992. "The Voyages of Discovery and the Loss and Rediscovery of the 'Other's' Knowledge." *Impact of Science on Society,* no. 167, 241–64.
————. Forthcoming. *Mining Civilizational Knowledge.*
Gould, Stephen Jay. 1981. *The Mismeasure of Man.* New York: Norton.
Griffin, Susan. 1978. *Woman and Nature: The Roaring inside Her.* New York: Harper and Row.
Hacking, Ian. 1983. *Representing and Intervening.* Cambridge: Cambridge University Press.
————. 1996. "The Disunities of the Sciences." In *The Disunity of Science,* ed. Peter Galison and David J. Stump. Stanford: Stanford University Press.
Haraway, Donna. 1989. *Primate Visions: Gender, Race, and Nature in the World of Modern Science.* New York: Routledge.
————. 1991. *Simians, Cyborgs, and Women: The Reinvention of Nature.* New York: Routledge.
————. 1991. "Situated Knowledges: The Science Question in Feminism and the Privilege of Partial Perspectives." In *Simians, Cyborgs, and Women.* New York: Routledge.
————. (Forthcoming). "Universal Donors in a Vampire Culture: Twentieth Century Biological Kinship Categories."
————. 1997. *Modest Witness@Second_Millennium.FemaleMan© Meets_Onco-Mouse™: Feminism and Technoscience.* New York: Routledge.
Harcourt, Wendy, ed. 1994. *Feminist Perspectives on Sustainable Development.* London: Zed.
Harding, Sandra, ed. 1976. *Can Theories Be Refuted? Essays on the Duhem-Quine Thesis.* Dordrecht: Reidel.
————. 1983. "Why Has the Sex/Gender System Become Visible Only Now?" In *Discovering Reality,* ed. Sandra Harding and Merrill Hintikka. Dordrecht: Reidel.
————. 1986. *The Science Question in Feminism.* Ithaca: Cornell University Press.
————, ed. 1987. *Feminism and Methodology: Social Science Issues.* Bloomington: Indiana University Press.
————. 1991. *Whose Science? Whose Knowledge?* Ithaca: Cornell University Press.
————. 1992a. "After the Neutrality Ideal: Science, Politics and 'Strong Objectivity.'" *Social Research* 59. 567–87; reprinted in *The Politics of Western Science,* ed. Margaret C. Jacob. Atlantic Highlands, N.J.: Humanities, 1993.
————. 1992b. "Rethinking Standpoint Epistemology." In *Feminist Epistemologies,* ed. L. Alcoff and E. Potter. New York: Routledge.

———, ed. 1993. *The "Racial" Economy of Science: Toward a Democratic Future.* Bloomington: Indiana University Press.

———. 1994. "Is Science Multicultural? Challenges, Resources, Opportunities, Uncertainties." In *Configurations* 2:2, and in *Multiculturalism: A Reader,* ed. David Theo Goldberg. London: Blackwell.

———. 1995a. "Can Feminist Thought Make Economics More Objective?" *Feminist Economics* 1:1. 7–32.

———. 1995b. "'Strong Objectivity': A Response to the New Objectivity Question." In *Synthese* 104:3. 1–19.

———. 1996a. "How Social Disadvantage Creates Epistemic Advantage." In *Social Theory and Sociology,* ed. Stephen Turner. New York: Blackwell.

———. 1996b. "Multicultural and Global Feminist Philosophies of Science: Resources and Challenges." In *Feminism, Science, and the Philosophy of Science,* ed. Lynn Hankinson Nelson and Jack Nelson. Dordrecht: Kluwer.

———. 1997a. "Women's Standpoints on Nature: What Makes Them Possible?" *Osiris* 12. 1–15.

———. 1997b. "Can Men Be Subjects of Feminist Theory?" In *Men Doing Feminism,* ed. Tom Digby. New York: Routledge.

———. 1998. "Are Truth Claims Dysfunctional?" In *Philosophy of Language: The Big Questions,* ed. Andrea Nye. New York: Blackwell.

———, and Merrill Hintikka, eds. 1983. *Discovering Reality: Feminist Perspectives on Epistemology, Metaphysics, Methodology, and Philosophy of Science.* Dordrecht: Reidel, 1983.

———, and Jean O'Barr, eds. 1987. *Sex and Scientific Inquiry.* Chicago: University of Chicago Press, 1987.

———, and Elizabeth McGregor. 1996. "The Gender Dimension of Science and Technology." In *UNESCO World Science Report,* ed. Howard J. Moore. Paris: UNESCO.

Hartsock, Nancy. 1983. "The Feminist Standpoint: Developing the Ground for a Specifically Feminist Historical Materialism." In *Discovering Reality: Feminist Perspectives on Epistemology, Metaphysics, Methodology, and Philosophy of Science,* ed. Sandra Harding and Merrill Hintikka. Dordrecht: Reidel/Kluwer.

Hayles, N. Katherine. 1992. "Gender Encoding in Fluid Mechanics: Masculine Channels and Feminine Flows." *Differences* 4:2. 16–44.

———. 1993. "Constrained Constructivism: Locating Scientific Inquiry in the Theater of Representation." In *Realism and Representation,* ed. George Levine. Madison: University of Wisconsin Press.

Headrick, Daniel R., ed. 1981. *The Tools of Empire: Technology and European Imperialism in the Nineteenth Century.* New York: Oxford University Press.

Hennessey, Rosemary. 1993. *Feminist Materialism and the Politics of Discourse.* New York: Routledge.

Hess, David J. 1995. *Science and Technology in a Multicultural World: The Cultural Politics of Facts and Artifacts.* New York: Columbia University Press.

Hesse, Mary. 1966. *Models and Analogies in Science.* Notre Dame: University of Notre Dame Press.

Hessen, Boris. 1970. *The Economic Roots of Newton's Principia.* New York: Howard Fertig.

hooks, bell. 1981. *Ain't I a Woman? Black Women and Feminism.* Boston: South End Press.

——. 1983. *Feminist Theory from Margin to Center.* Boston: South End Press.

Horton, Robin. 1967. "African Traditional Thought and Western Science," pts. 1 and 2. *Africa* 37.

Hsu, Elizabeth. 1992. "The Reception of Western Medicine in China: Examples from Yunan." In *Science and Empires,* ed. Petitjean et al.

Hubbard, Ruth. 1990. *The Politics of Women's Biology.* New Brunswick: Rutgers University.

——, M. S. Henifin, and Barbara Fried, eds. 1982. *Biological Woman: The Convenient Myth.* Cambridge, Mass.: Schenkman.

Hull, Gloria, Patricia Bell Scott, and Barbara Smith. 1982. *But Some of Us Are Brave: Black Women's Studies.* Old Westbury: Feminist Press.

Jaggar, Alison. 1983. "Chapter 11." *Feminist Politics and Human Nature.* Totowa, N.J.: Rowman and Allenheld.

——. 1989. "Love and Knowledge: Emotion in Feminist Epistemology." In *Gender, Body/Knowledge: Feminist Reconstructions of Being and Knowing,* ed. Alison Jaggar and Susan Bordo. New Brunswick: Rutgers University Press.

——, and Paula Rothenberg. 1993. *Feminist Frameworks.* New York: McGraw Hill.

James, C. L. R. 1963. *The Black Jacobins,* 2d ed. rev. New York: Vintage.

Jameson, Fredric. 1988. "'History and Class Consciousness' as an Unfinished Project." *Rethinking Marxism* 1. 49–72.

Joseph, George Gheverghese. 1991. *The Crest of the Peacock: Non-European Roots of Mathematics.* New York: I. B. Tauris.

Kanter, R. M. 1977. *Men and Women of the Corporation.* New York: Basic.

Kaptchuk, Ted J. 1983. *The Web That Has No Weaver: Understanding Chinese Medicine.* New York: Congdon and Weed.

Keita, Lacinay. 1977–78. "African Philosophical Systems: A Rational Reconstruction." *Philosophical Forum* 9:2–3.

Keller, Evelyn Fox. 1983a. *A Feeling for the Organism.* San Francisco: Freeman.

——. 1983b. "Gender and Science." In *Discovering Reality,* ed. Sandra Harding and Merrill Hintikka. Dordrecht: Reidel/Kluwer.

——. 1984. *Reflections on Gender and Science.* New Haven: Yale University Press.

——. 1992. *Secrets of Life, Secrets of Death: Essays on Language, Gender, and Science.* New York: Routledge.

Khor, Kok Peng. 1988. "Science and Development: Underdeveloping the Third World." In *The Revenge of Athena,* ed. Z. Sardar. London: Mansell.

Kline, Morris. 1980. *Mathematics: The Loss of Certainty.* New York: Oxford.

Kochhar, R. K. 1992–93. "Science in British India," pts. I and II. *Current Science* 63:11. 689–94; 64, 55–62 (India).

Krishna, V. V. 1992. "The Colonial 'Model' and the Emergence of National Science in India: 1876–1920." In *Science and Empires,* ed. Petitjean et al.

Kuhn, Thomas S. (1962). 1970. *The Structure of Scientific Revolutions,* 2d ed. Chicago: University of Chicago Press.

Kumar, Deepak. 1992. "Problems in Science Administration: A Study of the Scientific Surveys in British India, 1757–1900." In *Science and Empires,* ed. Petitjean et al.

————. 1991. *Science and Empire: Essays in Indian Context (1700–1947)*. Delhi, India: Anamika Prakashan and National Institute of Science, Technology, and Development.

Lach, Donald F. 1977. *Asia in the Making of Europe.* Vol. 2. Chicago: University of Chicago Press.

Lakatos, Imre, and Alan Musgrave. 1970. *Criticism and the Growth of Knowledge.* Cambridge: Cambridge University Press.

Lambert, Helen. 1987. "Biology and Equality: A Perspective on Sex Differences." In *Sex and Scientific Inquiry,* ed. Sandra Harding and Jean O'Barr. Chicago: University of Chicago Press.

Latour, Bruno. 1987. *Science in Action.* Cambridge: Harvard University Press.

————. 1988. *The Pasteurization of France.* Cambridge: Harvard University Press.

————. 1993. *We Have Never Been Modern.* Trans. Catherine Porter. Cambridge: Harvard University Press.

————, and Steve Woolgar. 1979. *Laboratory Life: The Social Construction of Scientific Facts.* Beverly Hills, Calif.: Sage.

Lennon, Kathleen, and Margaret Whitford. 1993. *Knowing the Difference: Feminist Perspectives in Epistemology.* New York: Routledge.

Li, C. P. 1977. "Chinese Herbal Medicine: Recent Experimental Studies, Clinical Applications and Pharmacognosy of Certain Herbs." In *A Barefoot Doctor's Manual,* rev. ed., Revolutionary Health Committee of Hunan Province. Seattle: Madrona.

Lloyd, Genevieve. 1984. *The Man of Reason: "Male" and "Female" in Western Philosophy.* Minneapolis: University of Minnesota Press.

————. 1993. "Maleness, Metaphor, and the 'Crisis' of Reason." In *A Mind of One's Own: Feminist Essays on Reason and Objectivity,* ed. Louise Antony and Charlotte Witt. Boulder: Westview Press.

Lloyd, Lisa. 1996. "Science and Anti-Science: Objectivity and Its Real Enemies." *Feminism, Science, and the Philosophy of Science,* ed. Lynn Hankinson Nelson and Jack Nelson. Boston: Kluwer.

Longino, Helen. 1990. *Science as Social Knowledge.* Princeton: Princeton University Press.

Lukacs, George. 1971. *History and Class Consciousness.* Cambridge, Mass.: MIT Press.

MacKinnon, Catharine. 1982. "Feminism, Marxism, Method, and the State: An Agenda for Theory." *Signs* 7:3.

MacLeod, Roy. 1987. "On Visiting the 'Moving Metropolis': Reflections on the Architecture of Imperial Science." In *Scientific Colonialism: A Cross-Cultural Comparison,* ed. Nathan Reingold and Marc Rothenberg. Washington, D.C.: Smithsonian Institution Press.

Martin, Emily. 1987. *The Woman in the Body: A Cultural Analysis of Reproduction.* Boston: Beacon.

————. 1994. *Flexible Bodies.* Boston: Beacon.

————. 1991. "The Sperm and the Egg." *Signs* 16:3. 485–501.

McClellan, James E. 1992. *Colonialism and Science: Saint Domingue in the Old Regime.* Baltimore: Johns Hopkins University Press.

Megill, Alan. 1992. "Rethinking Objectivity." In *Annals of Scholarship* 8:3, ed. Alan Megill.

Merchant, Carolyn. 1980. *The Death of Nature: Women, Ecology and the Scientific Revolution.* New York: Harper and Row.

Mies, Maria. 1986. *Patriarchy and Accumulation on a World Scale: Women in the International Division of Labor.* Atlantic Highlands, N.J.: Zed.

Mills, A. J., and P. Tancred, eds. 1991. *Gendering Organizational Theory.* London: Sage.

Mishra, Vijay, and Bob Hodge. 1994. "What Is Post(-) Colonialism?" In *Colonial Discourse and Post-colonial Theory,* ed. P. Williams and L. Chrisman. New York: Columbia University Press.

Moraze, Charles, ed. 1979. *Science and the Factors of Inequality.* Paris: UNESCO.

Moulton, Janice. 1983. "A Paradigm of Philosophy: The Adversary Method." In *Discovering Reality,* ed. Sandra Harding and Merrill Hintikka. Dordrecht: Reidel.

Nandy, Ashis, ed. 1990. *Science, Hegemony, and Violence: A Requiem for Modernity.* Delhi: Oxford.

Nasr, Seyyed Hossein. 1988. "Islamic Science, Western Science: Common Heritage, Diverse Destinies." In *The Revenge of Athena,* ed. Z. Sardar. London: Mansell.

National Academy of Sciences. 1989. *On Being a Scientist.* Washington, D.C.: National Academy Press.

Needham, Joseph. 1954 ff. *Science and Civilisation in China.* 7 vols. Cambridge: Cambridge University Press.

———. 1969. *The Grand Titration: Science and Society in East and West.* Toronto: University of Toronto Press.

Nelkin, Dorothy. 1987. *Selling Science: How the Press Covers Science and Technology.* New York: W. H. Freeman.

Nelson, David, George Gheverghese Joseph, and Julian Williams. 1993. *Multicultural Mathematics.* New York: Oxford University Press.

Nelson, Lynn Hankinson. 1990. *Who Knows.* Philadelphia: Temple University Press.

Noble, David. 1992. *A World without Women: The Christian Clerical Culture of Western Science.* New York: Knopf.

———. 1995. *The Religion of Technology.* New York: Knopf.

Novick, Peter. 1988. *That Noble Dream: The "Objectivity Question" and the American Historical Profession.* Cambridge: Cambridge University Press.

Nye, Robert A. 1993. *Masculinity and Male Codes of Honor in Modern France.* New York: Oxford.

———. 1995. "Science and Medicine as Masculine Fields of Honor." *Osiris* 12.

Okruhlik, Kathleen, et al. 1989. "Feminist Critiques of Science: The Epistemological and Methodological Literature." *Women's Studies International Forum* 12:3. 379–88.

Pacey, Arnold. 1990. *Technology in World Civilization: A Thousand-Year History.* Cambridge, Mass.: MIT Press.

Patterson, Tom. 1997a. "Nature: The Shadow of Civilization." Forthcoming.

———. 1997b. *Inventing Western Civilization.* New York: Monthly Review Press.

Petitjean, Patrick, et al., eds. 1992. *Science and Empires: Historical Studies about Scientific Development and European Expansion.* Dordrecht: Kluwer.

Pickering, Andrew. 1984. *Constructing Quarks*. Chicago: University of Chicago Press.

————. 1992a. "Objectivity and the Mangle of Practice." *Annals of Scholarship* 8:3, ed. Alan Megill.

————, ed. 1992b. *Science as Practice and Culture*. Chicago: University of Chicago Press.

Plumwood, Val. 1993. *Feminism and the Mastery of Nature*. New York: Routledge.

Polanco, Xavier. 1992. "World-Science: How Is the History of World-Science to Be Written?" In *Science and Empires*, ed. P. Petitjean et al. Dordrecht: Kluwer.

Powell, Arthur B., and Marilyn Frankenstein, eds. 1997. *Ethnomathematics: Challenging Eurocentrism in Mathematics Education*. Albany: State University of New York Press.

Proctor, Robert. 1988. *Racial Hygiene: Medicine under the Nazis*. Cambridge: Harvard University Press.

————. 1991. *Value-Free Science? Purity and Power in Modern Knowledge*. Cambridge: Harvard University Press.

————. 1995. *Cancer Wars: How Politics Shapes What We Know and Don't Know about Cancer*. Boston: Basic.

Quine, W. V. O. 1953. "Two Dogmas of Empiricism." In *From a Logical Point of View*. Cambridge: Harvard University Press.

Reingold, Nathan, and Marc Rothenberg, eds. 1987. *Scientific Colonialism: Cross-Cultural Comparisons*. Washington D.C.: Smithsonian Institution Press.

Restivo, Sal. 1988. "Modern Science as a Social Problem." *Social Problems* 35:3.

————. 1992. *Mathematics in Society and History: Sociological Inquiries*. Dordrecht: Kluwer.

————, Jean Paul Van Bendegem, and Roland Rischer, eds. 1993. *Math Worlds: Philosophical and Social Studies of Mathematics and Mathematics Education*. Albany: State University of New York Press.

Rodney, Walter. 1982. *How Europe Underdeveloped Africa*. Washington D.C.: Howard University Press.

Ronan, Colin A. 1973. *Lost Discoveries: The Forgotten Science of the Ancient World*. London: MacDonald.

Roof, Judith, and Robyn Wiegman, eds. 1995. *Who Can Speak? Authority and Critical Identity*. Urbana: University of Illinois Press.

Rooney, Phyllis. 1994. "Recent Work in Feminist Discussions of Reason." *American Philosophical Quarterly* 31:1. 1–21.

Rorty, Richard. 1979. *Philosophy and the Mirror of Nature*. Princeton: Princeton University Press.

Rose, Hilary. 1983. "Hand, Brain, and Heart: A Feminist Epistemology for the Natural Sciences." *Signs* 9:1.

————. 1984. *Love, Power, and Knowledge*. Bloomington: Indiana University Press.

————, and Steven Rose. 1976. "The Incorporation of Science." In *The Political Economy of Science: Ideology of/in the Natural Sciences*, ed. H. Rose and S. Rose. London: Macmillan.

Rosser, Sue. 1992. *Biology and Feminism*. New York: Twayne.

Rossiter, Margaret. 1982. *Women Scientists in America: Struggles and Strategies to 1940*. Baltimore: Johns Hopkins University Press.

————. 1995. *Women Scientists in America: Before Affirmative Action*. Baltimore: Johns Hopkins University Press.

Rouse, Joseph. 1987. *Knowledge and Power: Toward a Political Philosophy of Science*. Ithaca: Cornell University Press.

————. 1993. "What Are Cultural Studies of Scientific Knowledge?" *Configurations* 1:1. 1–22.

————. 1996. "Feminism and the Social Construction of Scientific Knowledge." In *Feminism, Science, and the Philosophy of Science*, ed. Lynn Hankinson Nelson and Jack Nelson. Dordrecht: Kluwer.

Roy, Rustum. 1985, 1986a, 1986b, 1987. "Letters." *Physics Today* 38:9, 9–11; 39:2, 15, 96; 39:4, 81–84; 40:2, 13.

Ruddick, Sara. 1989. *Maternal Thinking*. Boston: Beacon.

Sabra, I. A. 1976. "The Scientific Enterprise." In *The World of Islam*, ed. B. Lewis. London: Thames and Hudson.

Sachs, Wolfgang, ed. 1992. *The Development Dictionary: A Guide to Knowledge as Power*. Atlantic Highlands, N.J.: Zed.

Said, Edward. 1978. *Orientalism*. New York: Pantheon.

Sardar, Ziauddin, ed. 1988. *The Revenge of Athena: Science, Exploitation, and the Third World*. London: Mansell.

Sayers, Janet. 1982. *Biological Politics: Feminist and Anti-Feminist Perspectives*. New York: Tavistock.

Scheurich, James Joseph, and Michelle D. Young. 1997. "Coloring Epistemologies: Are Our Research Epistemologies Racially Biased?" *Educational Researcher* 26:3. 4–16.

Schiebinger, Londa. 1989. *The Mind Has No Sex? Women in the Origins of Modern Science*. Cambridge: Harvard University Press.

————. 1993. *Nature's Body: Gender in the Making of Modern Science*. Boston: Beacon.

Schuster, John A., and Richard R. Yeo, eds. 1986. *The Politics and Rhetoric of Scientific Method: Historical Studies*. Dordrecht: Reidel.

Seager, Joni. 1993. *Earth Follies: Coming to Feminist Terms with the Global Environmental Crisis*. New York: Routledge.

Sen, Gita, and Caren Grown. 1987. *Development Crises and Alternative Visions: Third World Women's Perspectives*. New York: Monthly Review Press.

Shapin, Steven. 1994. *A Social History of Truth*. Chicago: University of Chicago Press.

————, and Simon Shaffer. 1985. *Leviathan and the Air Pump*. Princeton: Princeton University Press.

Shiva, Vandana. 1989. *Staying Alive: Women, Ecology, and Development*. London: Zed.

Smith, Dorothy E. 1987. *The Everyday World as Problematic: A Sociology for Women*. Boston: Northeastern University Press.

————. 1990a. *The Conceptual Practices of Power: A Feminist Sociology of Knowledge*. Boston: Northeastern University Press.

————. 1990b. *Texts, Facts, and Femininity: Exploring the Relations of Ruling*. New York: Routledge.

Snow, C. P. (1959) 1964. *The Two Cultures: And a Second Look*. Cambridge: Cambridge University Press.

Sohn-Rethel, Alfred. 1978 *Intellectual and Manual Labor.* London: Macmillan.

Spanier, Bonnie B. 1995. *Im/partial Science: Gender Ideology in Molecular Biology.* Bloomington: Indiana University Press.

Sparr, Pamela, ed. 1994. *Mortgaging Women's Lives: Feminist Critiques of Structural Adjustment.* London: Zed.

Spivak, Gayatri. 1987. *In Other Worlds: Essays in Cultural Politics.* New York: Methuen.

Stepan, Nancy Leys. 1986. "Race and Gender: The Role of Analogy in Science." *Isis* 77.

Suppes, Patrick. 1978. "The Plurality of Science." *PSA 1978,* vol. 2, ed. P. Asquith and I. Hacking. East Lansing: Philosophy of Science Association.

Third World Network. 1988. *Modern Science in Crisis: A Third World Response.* Penang, Malaysia: Third World Network.

Tickner, J. Ann. 1992. *Gender in International Relations: Feminist Perspectives on Achieving Global Security.* New York: Columbia University Press.

Tiffany, Sharon, and Kathleen J. Adams. 1985. *The Wild Woman: An Inquiry into the Anthropology of an Idea.* Cambridge, Mass.: Schenkman.

Tobach, Ethel, and Betty Rosoff, eds. 1978, 1979, 1981, 1984. *Genes and Gender,* vols. 1–4. New York: Gordian.

Todorov, Tzvetan. 1984. *The Conquest of America: The Question of the Other.* Trans. Richard Howard. New York: Harper and Row.

Traweek, Sharon. 1988, *Beamtimes and Lifetimes.* Cambridge, Mass.: MIT Press.

———. 1993. "An Introduction to Cultural, Gender, and Social Studies of Sciences and Technologies." *Culture, Medicine, and Psychiatry* 17. 3–25.

———. Forthcoming. *Tinkering with Cultural Studies of Science.*

———. Forthcoming. *Big Science in Japan.*

Tuana, Nancy, ed. 1989. *Feminism and Science.* Bloomington: Indiana University Press.

Turnbull, David. 1993. "Local Knowledge and Comparative Scientific Traditions." *Knowledge and Policy* 6:3/4. 29–54.

United Nations Commission on Science and Technology for Development Gender Working Group. 1995. *Missing Links: Gender Equity in Science and Technology for Development.* Ottawa: International Development Research Centre.

Van den Daele, W. 1977. "The Social Construction of Science." In *The Social Production of Scientific Knowledge,* ed. E. Mendelsohn, P. Weingart, and R. Whitley. Dordrecht: Reidel.

Van Fraassen, Bas, and Jill Sigman. 1993. "Interpretation in Science and in the Arts." In *Realism and Representation,* ed. George Levine. Madison: University of Wisconsin Press.

Wajcman, Judy. 1991. *Feminism Confronts Technology.* University Park: Pennsylvania State University Press.

Wallerstein, Immanuel. 1974. *The Modern World-System.* Vol. 1. New York: Academic.

Watson-Verran, Helen, and David Turnbull. 1995. "Science and Other Indigenous Knowledge Systems." In *Handbook of Science and Technology Studies,* ed. S. Jasanoff, G. Markle, T. Pinch, and and J. Petersen. Thousand Oaks, Calif.: Sage. 115–39.

Weatherford, Jack McIver. 1988. *Indian Givers: What the Native Americans Gave to the World.* New York: Crown.

Weedon, Chris. 1987. *Feminist Practice and Poststructuralist Theory.* Cambridge, Mass.: Blackwell.

Wellman, David. 1977. *Portraits of White Racism.* New York: Cambridge University Press.

Williams, Eric. 1944. *Capitalism and Slavery.* Chapel Hill: University of North Carolina Press.

Williams, Patrick, and Laura Chrisman, eds. 1994. *Colonial Discourse and Post-Colonial Theory.* New York: Columbia University Press.

Winner, Langdon. 1996. "The Gloves Come Off: Shattered Alliances in Science and Technology Studies." In *Social Text* 14:1. 81–91.

Wiredu, J. E. 1979. "How Not to Compare African Thought with Western Thought." In *African Philosophy*, 2d ed., ed. Richard A. Wright. Washington, D.C.: University Press of America.

Wolf, Eric. 1984. *Europe and the People without a History.* Berkeley: University of California Press.

"Women in Legal Education—Pedagogy, Law, Theory, and Practice." 1988. *Journal of Legal Education,* 38:1 and 2. March/June.

Woolgar, Steve. 1988. *Science: The Very Idea.* New York: Tavistock.

———, ed. 1988. *Knowledge and Reflexivity.* Beverly Hills: Sage.

Worsley, Peter. 1967. *The Third World,* 2d ed. London: Weidenfeld and Nicolson.

Yates, Frances. 1969. *Giordano Bruno and the Hermetic Tradition.* New York: Vintage.

Zilsel, Edgar. 1942. "The Sociological Roots of Science." *American Journal of Sociology* 47.

INDEX

SANDRA HARDING, Professor of Education and Women's Studies at UCLA, is the author of *Whose Science? Whose Knowledge? Thinking from Women's Lives* and *The Science Question in Feminism* (winner of the Jessie Bernard Award of the American Sociological Association). She is the editor of *Feminism and Methodology: Social Science Issues* and *The "Racial" Economy of Science: Toward a Democratic Future,* and coeditor of *Discovering Reality: Feminist Perspectives on Epistemology, Metaphysics, Methodology, and Philosophy of Science* (with Merrill Hintikka) and of *Sex and Scientific Inquiry* (with Jean O'Barr).